Written for the author's grandchildren this enjoyable little book has a light, witty tone that makes it easy to digest.

...will keep you reading until you understand the special theory of relativity - plus a few more physics oddities the author throws in.

The *Newspaper of the Central Coast.*
TRIBUNE
San Luis Obispo, California.

"Everyone has heard of $E=mc^2$," Knight explained. "I wrote this book to show it's not very complicated."

....decided to spend some time writing for his grandchildren, but somehow it got out of hand.

The Times Press Recorder.
San Luis Obispo County.

An Introduction to…

The (1905) Special Theory of Relativity.

How something strange about the speed of light led to the discovery that mass can be changed into energy!

An explanation of special relativity for those who'd like to know more about it but thought it was too complicated.

Published by Triumph Books.

198 Squire Canyon Road,
San Luis Obispo, CA 93401-8001.
(805) 627-1778
tri-books@peoplepc.com

Library of Congress Control Number: 2004195044

ISBN-10: 9749544-2-X
ISBN-13: 978-0-9749544-2-4

Third Edition.

Printed in the United States of America.

Also published in The Sub-continent of India, including
India, Sri-Lanka, Nepal, Bangladesh and Pakistan
by

Alchemy

An imprint of Mehras.
15, Bankim Chatterjee Street, 2nd floor, Kolkata-700 073.
4767/23, Pratap Street, Darya Ganj, New Delhi -110 002.
G-1 and 2, Ghaswala Tower, P.G. Solanki Path, off Lamington Road,
Mumbai -400 007.

This book was written for the author's grandchildren, to encourage an interest in science.

For Alex, Candice, Clive, Devon, Jordan, Kelly-Ann, Roman, Taylor and Zach.

Horatio:

Oh day and night, but this is wondrous strange!

Hamlet:

And therefore as a stranger give it welcome.
There are more things in heaven and earth, Horatio,
Than are dreamt of in your philosophy.

FOREWORD.

A hundred years since Einstein's annus mirabilis.*

This little book was written by an average sort of person, in the hope that the subject may be more easily understood than would have been the case had it been written by someone who was very clever.

There are two theories of relativity. These are the *special* theory, which Einstein published in 1905, and the somewhat more complex *general* theory which came about ten years later.

Special relativity was the culmination of work done over the years by a number of people, and by 1905 was probably inevitable. Einstein was not the only one working on the idea, and had he not published his paper when he did there's little doubt someone else would have produced a similar document before long. Einstein agreed this was probably true. The facts had been gathered and all that was left then was for some exceptionally clever person to put it all together. That person turned out to be Albert Einstein.

General relativity is a bit more complicated, and in developing this theory it is generally recognized that Einstein was ahead of his time. We'll be touching briefly on general relativity in appendices (2) and (4) but otherwise it is beyond the scope of this book.

When anyone mentions special relativity most people throw up their hands and say it's much too difficult to understand.

It's true that many books on the subject are hard to follow. Others present the subject in simpler terms but don't go into the mathematics in any depth, merely presenting the various equations without showing how they were derived.

* 1905, his miracle year, as it has sometimes been called. Einstein actually submitted a total of no less than six papers that year, all of which were published. (More on this on page 11.)

In this book an attempt has been made to explain the subject in detail, showing how the equations were derived and the reasoning behind them, and yet do so in a manner that is easy to understand.

After all, it's not as though we're inventing special relativity. All we're going to be doing in this book is looking at something that was figured out by someone else a hundred years ago.

A discussion of special relativity has to involve some math. if the subject is to be covered properly, though most of it is no more than high school level. (Depends on what they teach you in high school these days.) But if you do find some of it hard going then just skip over it. It is possible to gain an understanding of special relativity without having to follow all the mathematics.

Part one of the book is fairly straight forward. Part two is a bit more complicated, but shouldn't be too difficult to follow if the reader just takes it one step at a time. It may be a challenge, but then people should be challenged. That was certainly the belief in this author's younger days.

If you come to something you can't follow, go back over it again before moving on to the next step. Don't allow yourself to be deterred by a temporary obstacle. Persevere, it's not that hard. Your brain is more than adequate for the task. The answers are out there waiting to be uncovered by the curious mind. With a little effort on your part you'll be able to lift a tiny corner of the veil beneath which lay concealed the mysteries of the universe.

It all started when scientists discovered that the speed of light was always the same, no matter how the source of the light (or the observer) was moving, and realized they had a problem!

A railroad coach is traveling along the track at a high rate of speed. There is a door at each end of the coach. To test the doors a man inside the coach pushes a button to open the doors. He sees both doors open simultaneously.

Another man sitting on an embankment alongside the track sees the coach rush past. He sees one door open first, then the other door open a moment later!

There are two twin brothers. One brother is an astronaut. The astronaut brother sets out on a long journey through outer space at very high speed. When he gets back he finds that, despite their both having been born on the same day, his twin brother who'd stayed behind on earth was now much older than he.

A man sees an iron bar whiz past him at high speed. He'd previously had an opportunity to measure this same iron bar and found it to be exactly one foot long. But now as the iron bar is going by the man sees that it's only a little over ten and three-eighths inches long.

You push something to get it moving. You push it very hard, and keep on pushing it so that it goes faster and faster. But after a while (it's going very fast now) it becomes hard to make it go any faster. As you continue pushing, instead of going faster you notice it's now getting *heavier* and *heavier*.

How are these things possible?

In this little book we'll explain that…

It's all a matter of relativity.

This study of Special Relativity will consist of two parts:

Part One is about how we got to the idea of Special Relativity,

and

Part Two describes some of the funny things that result from this discovery.

Part One.

How Einstein (with the aid of men like Galileo, Poincare, Lorentz and Newton) developed the Special Theory of Relativity.

THE SPECIAL THEORY OF RELATIVITY.

In 1905, in Volume 17, fourth series, of the German scientific journal *Annalen der Physik*, there appeared three papers, each one of considerable importance, submitted by a twenty-six year old clerk (technical examiner third class) in the Swiss patents office by the name of Albert Einstein.

The first of the three papers, submitted in March of that year, on the photo-electric effect, was subsequently to earn Einstein a Nobel Prize. The second paper, which had been submitted on May 11th, concerned what is known as the Brownian Motion. But a third paper, submitted on June 30th and consisting of thirty pages, the one we're going to be discussing in this book, and which never did win him a prize (see page 99) introduced to the world what we all know today as the Special Theory of Relativity.

Two months later Einstein published a fourth paper, this one of only three pages, in which he showed the relationship between mass and energy.*

Einstein said weird things happen near the speed of light!

The original paper on special relativity was actually entitled *Zur Elektrodynamik bewegter Korper* (on the electrodynamics of moving bodies). The other paper, the one of just three pages, posed the question *Ist die Tragheit eines Korpers von seinem Energienhalt abhangig* (does a body's inertia depend on its energy content).

The greatest scientist who ever lived, Sir Isaac Newton, once said that if he'd seen farther it was because he'd been standing on the shoulders of others. (Actually he used the word giants.)

Like Newton, Einstein was standing on the shoulders of those who'd gone before him. So for a proper understanding of special relativity we must first look at the events that led up to it. We need to review the achievements of those earlier figures upon whose shoulders Einstein could be said to have been standing.

* He'd also been working on his doctoral thesis in his spare time, a thirty-one page dissertation he submitted on July 20th of that year. So Einstein must have been a busy boy around that time!

By the end of the nineteenth century scientists were starting to piece together certain known facts, and realized that some of them didn't make sense. It was the speed of light that bothered them mostly. Of all those working on the problem it was Einstein who was the first to develop and publish a satisfactory explanation.

This new idea of special relativity attracted little notice at first, and what comment it did provoke at the time was mostly skeptical, if not scornful. It seemed to be of little practical value.

Then in 1916 Einstein published his theory of general relativity, extending the concept to encompass the universe as a whole! Still, as we told you in the foreword, apart from being briefly touched on in appendices (2) & (4) *general* relativity is not a subject with which we shall be concerning ourselves in this little book.

One of those upon whom Einstein relied was Galileo, who lived from 1564 to 1642. Galileo died on Christmas Day 1642, so most sources say, the same day that Sir Isaac Newton was born,* although England was still on the old Julian calendar at that time, which meant that by the time Newton was born it was already Jan. 4th in Catholic Italy and Galileo had actually been dead for ten days.**

Galileo was an Italian scientist who left his mark in a number of fields, most notably, in 1609, as the first to use the telescope to study the night sky, though he didn't invent the telescope as is sometimes thought.

Well, what's it going to be tonight, gentlemen? The great nebula in Orion, or the blonde in the apartment across the street?

Galileo also supplied a telescope to the Doge of Venice for a huge sum of money plus a lifetime professorship. He had recognized the usefulness of such an instrument to warships of the Doge's navy.

* He didn't become *"Sir"* Isaac until later, of course.

** Some give the year of Newton's birth as 1643. This would be correct by the modern Gregorian calendar which Protestant Britain, along with her American colonies, eventually adopted in 1752. By then the change-over involved the losing of *eleven* days.

Galileo's celestial observations led him to the conclusion that the earth was not at the center of the universe.*

At which the Pope suggested the learned gentleman may want to give the matter a little more thought or he'd find himself at the center of a barbeque in the town square.

The Pope said that everything in the universe went around the earth.

However, here we're only going to be talking about Galileo's studies of things that are *moving*.

In part one we'll only be dealing with things that are moving *steadily and in a straight line*. We won't be dealing with things that are speeding up, or slowing down, or with things that are turning in one direction** or another.

In fact it should be understood from the very start that *special relativity only applies if the bodies involved are moving at a steady speed and in a straight line.*

First we'll talk about the *Principle* of Relativity. No, not the *Theory* of Relativity (that came later). But some three hundred years before Albert Einstein appeared on the scene Galileo paved the way for special relativity by developing what he called the *Principle* of Relativity.

Incidentally, Galileo never minded letting people *think* that he'd invented the telescope, while it was really invented by someone else in Holland, who doesn't seem to have known what to do with it.

* With his new telescope Galileo had discovered the four largest of Jupiter's moons (Io, Europa, Ganymede and Calisto). *They were all going round Jupiter*. This meant that *not everything* in the universe went around *the earth*. It was in line with what Nicolaus Copernicus (1473-1543) had been saying before Galileo was even born!

** This is what people would normally say. But to engineers things don't turn in a *direction*. To us a direction implies a straight line. If a thing is turning we say it's turning in a *sense*. Like in a clockwise *sense*, or in a counter-clockwise *sense*. (Though unfortunately these days such distinctions are not always observed.)

CHAPTER ONE.

THE PRINCIPLE OF RELATIVITY.

A man is sitting by the side of a road. Near him is a tree. The man sees a car coming along the road towards him. It is traveling at 50 miles per hour.

What does this mean? It means the car is traveling at 50 miles per hour *as seen by* the man. Or, to put it another way, traveling at 50 miles per hour *relative to* the man sitting beside the road.

Suppose there's another car. The second car is behind the first car and is traveling at only 40 miles per hour. By this we mean 40 miles per hour *relative to* the man sitting by the side of the road.

This second car could also be said to be traveling at 40 miles per hour *relative to* the tree, because the tree, like the man (at least as far as anyone can tell) isn't moving at all.

Now suppose the man in the first car looks back at the second car. The second car is falling farther and farther behind him because it's only going 40 miles per hour while he's going 50.

If the man in the first car takes his mind off his driving for a moment and concentrates all his attention on the car that is behind him, then what does he see? He sees that he is pulling away from the second car, because the second car (the one that's behind him) is going 10 miles per hour slower than he is.

As seen by (or *relative to*) the man in the first car, the second car is traveling at 10 miles per hour not 40. *The distance between the two cars is changing at the rate of 10 miles per hour!*

Similarly, *relative to* the man in the second car, the *first* car is traveling, *not at 50* but at only 10 miles per hour. Each of the two cars, *relative to the other*, is traveling at 10 miles per hour.

All motions are *relative* not absolute! We have to rid ourselves of the idea that *anything* (like the earth, for example) can be *fixed!*

Whether a body is in motion or at rest, or how fast it is going, depends strictly on the point of view of the observer!

To sum up, if you're going to say a thing's moving, or how fast it's going, then you have to say what the motion is *relative to*.

Relative to the man sitting by the side of the road the first car is traveling at 50 miles per hour. But *relative to* the man in the second car it's only going 10 miles per hour.

Now we know that the first car is traveling at 50 miles per hour towards the man sitting by the side of the road. But Galileo said it's just as true to say that the man sitting by the side of the road is traveling at 50 miles per hour *towards the car.* (Actually, of course, there weren't any cars in Galileo's time.)

The *tree* too (Galileo would have said) *is traveling at 50 miles per hour towards the car!*

We think of the earth as being *fixed* and everything else moving *relative* to it. But Galileo says that all motion must be considered only as *relative* motion. *Whether one of the bodies we're studying happens to be stationary relative to the earth is of no importance!*

What we're considering here is simply a *car* and a *tree*. The car and the tree are drawing closer to each other. Forget about the earth! *The earth has nothing to do with it!*

Imagine for a moment that you could wave a wand and make everything disappear except for the car and the tree. Now all we see is a car and a tree drawing closer to each other! Relative to the tree the car is moving. *Relative to the car the tree is moving!* If *nothing is fixed* then the one is just as true to say as the other!

There's a story that Einstein was on a train in Switzerland. In about ten minutes the train was due to arrive at Zurich. A little old lady came up to Einstein and told him she thought his Theory of Relativity was absolutely marvelous, but unfortunately she didn't understand it. She asked Einstein if he'd be kind enough to explain it to her. Einstein was getting ready to leave the train at Zurich and was irritated. (Both Einstein and Newton are said to have been very unpleasant men.) "*Relativity,*" Einstein told the lady, "*means that in about ten minutes Zurich will be arriving at this train.*"

Let's take another case of relative motion. The man sitting by the side of the road sees a jet 'plane flying overhead. It is flying at 500 miles per hour (*relative to* the man).

In the cabin of the 'plane a flight attendant is fetching a drink for a passenger seated at the rear of the 'plane. She is walking towards the rear of the 'plane at about four miles per hour *relative to* the pilot and to the passengers who are seated. To the man sitting by the side of the road, how fast is the flight attendant traveling?

Well we know that the 'plane is traveling at 500 miles per hour relative to the man sitting by the side of the road. But what about the flight attendant? To the man sitting by the side of the road, how fast is *she* traveling?

The flight attendant is walking towards the rear of the 'plane, so to the man sitting by the side of the road she's traveling four miles an hour *slower* than everyone else on the 'plane.

Relative to the pilot, and to the passengers who are seated, the flight attendant is traveling at four miles per hour. But *relative to the man sitting by the side of the road* she is traveling at 496 miles per hour (the 500 miles per hour that the 'plane and everyone else on board is moving *forward,* less the four miles per hour that she is walking *in the opposite direction*).

Galileo said the speed of a thing will vary and depends on what it's being measured *relative to*. You can't say a thing is *moving* unless you say what it's moving *relative to*! To just say an object is *moving,* or to say it's *standing still,* is meaningless!

We're not quite finished talking about things that are moving yet. There's one more thing we have to consider. It has to do with things that occur in the same frame of reference. (There'll be more about frames of reference later.) In the meantime let's consider the following case.

A man and his wife have purchased tickets for a cruise, leaving from New York. They are now in their cabin. The ship is leaving harbor, moving at a steady speed and in a straight line, but they don't know this yet because the sea is calm that day, and because the noise of the ship's engines can't be heard from their cabin. The man says to his wife, "I wonder if we're moving yet, do you know?" His wife says she doesn't know.

The wife has an idea. She goes to the porthole and looks out. She's just in time to see the Statue of Liberty disappearing astern. A few minutes later and Monsieur Bartholdi's beckoning beacon is far behind them.* She tells her husband that the ship is moving!

There was no way to tell they were moving except by looking out of the porthole and observing some other body, and noting the ship's motion relative to that body. In this particular case the body happened to be the Statue of Liberty Enlightening the World. Or it could have been, say, a lighthouse!

That will be enough about relative motion. In the next chapter we're going to talk about waves.

* Did you know that the French had originally designed the Statue of Liberty as a lighthouse, to be erected at the northern entrance to the Suez Canal representing "*Progress Bringing The Light To Asia.*" The English having somehow gained control of the canal the statue's designer, a man by the name of Frederick Auguste Bartholdi (1834-1904) was left without a place to put it! So he said why not alter it a bit and give it to the Americans. (Fred didn't care *why* the thing got built, just so long as it got built!) It would help improve relations with their former allies, he claimed. For after helping the Americans apply the finger to King George, the French then made the mistake of backing the wrong side in the war between the states. This caused the victorious Union to favor the Prussians in the disastrous (for the French) Franco-Prussian war of 1870-71, and the French never really amounted to anything much again after that! This new plan being favorably received the recycled aid-to-navigation was ceremoniously unveiled (ten years too late for America's centennial celebrations) to welcome those of the huddled masses whose yearning to breath free was assuaged in October 1886. The writer followed not quite eighty years later, by a form of transport which, unfortunately, does not afford the wretched refuse from a foreign shore (as Emma Lazarus called us) the traditional view of the copper colossus upon arrival.

CHAPTER TWO.

ALL ABOUT WAVES.

Waves occur in water. Other places too. If you throw a stone into a swimming pool you'll see little waves spread out in circles from the point where the stone hit the water.

There's a line in Shakespeare where one of the actors* describes glory as "*...like a circle in the water, which never ceaseth to enlarge itself until, by broad spreading, it disperse to naught.*" **

The circles (waves) move outwards. You can see them! But the *water* doesn't move. A leaf floating on the water might bob up and down as the wave passes beneath it, but it will still be in the same place after the wave has gone. The water doesn't really *do* anything. But it has to *be* there. If you threw a stone into an empty swimming pool there wouldn't be any waves.

Sometimes waves travel through the air. Our ears detect these waves. *They are what we hear.* These waves have different lengths. High notes have short waves, and low notes have long waves

To fill up the same distance (the distance between the source of the sound and the listener) there has to be *more* of the short waves. Since the sound waves are all traveling at the same speed (and since there's more of them) the short waves will arrive at the listener's ear more frequently. Short waves are said to have a higher *frequency.*

If you hit the middle C note on a piano the sound waves will be generated at the rate of about 263 waves per second. If you hit the note one octave higher the waves will be *shorter,* and will leave the piano at a rate of about 526 waves per second. That's how frequently the sound waves will be produced. It is their *frequency.*

* We know her as Joan of Arc.

** King Henry VI- Part I. Act I, Scene iii.

We said that sound waves travel through the air. They can travel through other things too. But they have to travel through *something!* If you take an electric bell, like a door bell say, and put it inside a glass jar, when you push the button you'll hear the bell ring. A bit muffled, perhaps, because it's inside a glass jar. But you'll hear it.

Now if you take a pump and start to suck the air from the glass jar then the ringing of the bell will become weaker. Eventually, as you remove more and more air from the jar, you won't be able to hear the bell at all! You'll see through the glass that the hammer is banging against the bell, but you won't be able to *hear* it.

You can't have ripples in a swimming pool unless there's water in the pool. And you can't have sound waves unless there's some medium, like air, for them to travel through.

The speed of the sound wave depends on what it's traveling through. In air at sea level sound travels at about 761 miles per hour.* A bit slower at the top of a mountain, where the air is thinner, and a bit faster at low-lying places like Death Valley, say, which is below sea-level where the atmospheric pressure would be slightly higher. Or in one of those recompression chambers where divers sometimes have to spend time to avoid getting the bends caused by expansion of nitrogen bubbles in their blood. Because the air in these chambers is *compressed* sound travels significantly faster.

Let's just round it off and say sound travels at 760 miles per hour. But keep in mind that its speed actually depends on what it's traveling *through*, and on things like the temperature and pressure, and the motion of the medium through which it is traveling.

Can we get some service back here!

Now imagine you were crossing the Atlantic on the Concord which, until withdrawn from service in 2003, was a jet 'plane that traveled faster than sound. Suppose you wanted to call the flight attendant. If she was further forward in the aircraft than you were would she be able to hear you?

* In water it travels five times faster than this.

After all the Concord, together with the cabin flight attendant, is traveling at 1,350 miles per hour. The sound of your voice wouldn't be able to catch up with her, would it? Sound travels at only around 760 miles per hour and the flight attendant is moving at 1,350 miles per hour. That's much too fast for the sound to catch up with her.

Of course everyone knows that passengers on the Concord can (or *could*) be heard by people further forward in the 'plane. This is because the speed of a sound wave is *relative to the medium through which it's traveling.*

So to anyone on the ground, if they had any way of measuring the speed of the sound waves coming from the man's mouth, as they made their way to the ears of a flight attendant located further forward in the 'plane, then *relative to* a person on the ground *those particular sound waves* would be traveling at 2,110 miles an hour.

To other passengers though, the sound of the man's voice would be traveling at only the normal speed of about 760 miles per hour.

The speed of sound is always *relative to* the medium through which it's traveling. Remember the Concord, and everything in it, including the air in the cabin, was traveling at 1,350 miles per hour. And the sound of the man's voice was traveling *through the air* at 760 miles an hour.* So for a man on the ground *you have to add the speed of the sound through the air, and the speed of the air itself!*

Sound can be emitted at a certain wavelength, so at a certain pitch or note, yet be heard by a listener at a *different* pitch or note.

At a race car event, when a car is approaching you the noise from the engine will sound at a certain note. But as the car passes you and is moving away at high speed the note will suddenly drop to a lower pitch. This is called the Doppler Effect, named for Christian Johann Doppler (1803-1853) an Austrian physicist who first investigated this phenomenon.**

Light also travels in waves. The different colors of light have different wave lengths, so they have different *frequencies.*

* For the sake of simplicity this assumes the cabin is pressurized to atmospheric pressure at sea level. In real life it would be somewhat less than this, and the sound would actually travel a bit slower.

** For more on the Doppler Effect see appendix (9).

The longest light waves *that we can see* are the red ones, which have a wavelength of only about 0·00007 centimeters, or seven-tenths of a micron.* The shortest *that we can see* are the violet waves, with an even shorter length of only 0·00004 centimeters, or four-tenths of a micron.

Considering the virtually unlimited range of electromagnetic wavelengths in general (page 26) you'll note that all *visible* light lies within a very narrow band indeed.

Light waves longer then red are called infra-red. We can't see them. Light waves shorter than violet are called ultra-violet. We can't see them either.

All light waves regardless of whether they're long or short travel at the same speed.

If this were not so, then suppose we'd been watching the eclipse of some heavenly body by a dark neighbor, when the eclipse was over and the heavenly body became visible again the light would start to arrive here at different times according to color. This doesn't happen. No, all light travels at the same speed.

In the next chapter we're going to explore the question of light a little more thoroughly.

* A micron is a millionth of a meter.

CHAPTER THREE.

ALL ABOUT LIGHT.

Then God said "Let there be light;"
and there was light.
And God saw the light, that it was good;
and God divided the light from the darkness.

Genesis 1, verse 3.

Awake! for Morning in the Bowl of Night
Has flung the Stone that puts the Stars to Flight:
And Lo! The hunter of the East has caught
The Sultan's Turret in a Noose of Light.

The Ruba'iya't of Omar Khayya'm.
(Fitzgerald's original translation, opening stanza.)

How fast does light travel? Light travels so fast people used to think it was instantaneous. Then around the year 1676 the Danish scientist Ole (or Olaus) Christensen Roemer showed that light takes a definite *time* to travel. It takes one second for light to travel 225,000 kilometers, he said.

Because light travels so fast its speed was difficult to measure accurately with the equipment they had back then.

In November of 1972 a measurement put the speed of light at 299,792,456·2* meters per second (which is about 186,282·38 *miles* per second). To keep it simple we'll call it 186,000 miles per second.

Light travels so fast it can circle the earth about seven and a half times in one second. It takes only about one and a half seconds to get here from the moon. Light from the sun takes only a little over eight minutes to reach us from 93 million miles away.

But wait a moment, you say. You just told us on page 17 that if you're going to say how fast a thing's moving, then you have say what that speed is *relative to*. Otherwise it doesn't *mean* anything.

* The figure of 300 million meters per second used by amateur radio operators is close! The length of the wave, in meters, multiplied by the frequency, in megahertz, always comes out to 300. So if you know one of them you can easily find the other.

You're absolutely right. But we're going to skip that for now and come back to it later. Right now we're going to talk about how light gets here from the sun and the stars. Because if light is a *wave* then it has to have something to travel through.

But light from the sun and stars (particularly the stars, because they're so far away) has to travel through space, and space is *empty!* (It's so empty that a volume of space the size of the earth is estimated to contain a *mass* no greater than that of about a single grain of sand.)

Remember the bell inside the glass jar from which we'd sucked (almost) all the air? We couldn't *hear* the bell because there was no air for the sound to travel through. But we could *see* the hammer of the bell moving. Even though the *sound* couldn't reach us the *light* could! It didn't need the air to travel through!

Yet scientists were still convinced there had to be *something* for light to travel through and decided that space must be filled with an invisible substance they called *aether* (or ether). Scientists at the time referred to it as the luminiferous (light-transmitting) aether.

Even Einstein believed in aether at first. It filled all of space and didn't move. (They should have talked to Galileo about *that!*)

Light, it was said, traveled in the form of electromagnetic *waves* in this invisible yet ubiquitous medium which so far no-one had been able to find. And these electromagnetic waves traveled at 186,000 miles per second *relative to the aether*, that was the important thing.

Radio waves, being electromagnetic waves, must therefore also travel through this mysterious aether, it was assumed. In the early days of radio, when crystal sets and cats whiskers were giving way to thermionic *tubes** (in Britain they were called *valves*) this author remembers how you'd sometimes hear radio announcers say they were "*coming to you through the aether.*"

This is Station WXYZ coming to you through the aether.

Scientists found no evidence of aether, but were sure it was there and they'd find it if they just kept looking.

* This was long before the invention of transistors which, because of their small size, were originally developed for hearing aids.

Einstein wondered what would happen if he were able to fly through the aether *faster* than 186,000 miles per second.

Imagine Einstein is flying through the aether and approaching the speed of light. In front of him he's holding a mirror in which he can see his face.

Now he's flying through the aether. He's going faster and faster. Suddenly his face disappears from the mirror! What has happened?

What's happened is that Einstein has reached the speed of light. The light can no longer travel from Einstein's face to the mirror because the mirror he's holding out in front of him is going too fast for the light from his face to catch up with it. (Remember, the light is traveling at 186,000 miles per second *relative to the aether!*)

What's wrong with this? When Einstein's face disappears from the mirror he knows he's now reached the speed of light. He doesn't need to look at any outside body, like the Statue of Liberty or a lighthouse (page 18) to know that he's moving. This would violate Galileo's Principle of Relativity which we discussed in chapter one.

Einstein thought Galileo was probably right. He felt pretty sure he'd continue to see his face in the mirror no matter how fast he was going, and doubted there was any such thing as aether. But this still left unanswered the question of how light could travel as a wave if there was no medium for it to make waves in.

Meanwhile two Americans, Albert Michelson (1852-1931) and Edward Morley (1838-1923) had been conducting some experiments. They knew that in going round the sun the earth was moving through the aether at about 20 miles per second. If the aether didn't move (and that was the whole idea of aether) then it should be possible to detect the motion of the earth going through it. But Michelson and Morley couldn't detect the motion of the earth through the aether. Nor could anybody else! Which meant that in all likelihood it (the aether, that is) didn't exist!

Michelson, later to become the first American to win the Nobel prize for physics, was a Polish born Annapolis graduate.

Einstein learned of these experiments later but, as we said, by *that* time he'd come to pretty much the same conclusion anyway.

If a luminiferous aether didn't exist it occurred to the inquisitive young man that perhaps light didn't *always* have to be a wave. That maybe it could take on some other form that didn't *require* a medium for it to travel through! And if such were the case the same could apply to other things we always *thought* were waves.

In the next chapter we're going to examine further this question of whether light doesn't necessarily *always* have to travel as a wave. Meanwhile here's a list of things that, up until now, we've always thought of as being waves. Cosmic rays are not listed here since they're not really rays at all, but are the atomic nuclei of various elements such as iron or oxygen traveling at *close to*, but never quite able to achieve, the speed of light. Cosmic rays have high energy, though, even more energy than gamma rays. (See also page 36.)

There seem to be no limits to how long or how short these waves can be. From gamma rays on down. Radio waves can be longer than *ten million centimeters*, or more than *sixty miles!* On the other hand gamma rays can be less than a *billionth* of a centimeter! Nor do even these appear to be the extremes. It is thought that gamma rays can have wavelengths of less than a *trillionth* (10^{-12}) of a centimeter, while radio waves can stretch out virtually to infinity.

The shorter the wavelength the higher the frequency. As waves are continuous and all travel at the same speed then (just like sound waves) if they're *shorter* that means there has to be *more of them.*

The shorter the wavelength the more energy the wave has and so the more dangerous it is! Below is a table of electromagnetic waves arranged in the principal groups to which we have assigned them, starting with the shortest:

Gamma rays.	(Most dangerous of all.)
X-rays	(They're dangerous too.)
Ultraviolet rays.	(Don't stay in the sun too long.)
Visible light.	
Infrared rays.	
Microwaves.	(Can be dangerous when highly concentrated.)
Radio waves.	(The longest of these are pretty harmless.)

Because radio waves are long, and don't have much energy, the longest of them are unable to punch through the various layers of the ionosphere and instead bounce back to earth again. Usually they bounce off what are called the E, and the F_1 and F_2 layers. (At night there's no E layer, and only a single F layer.)

This is how they're able to be transmitted from one side of the earth and be received by someone else on the other side of the earth. They just keep bouncing off the ionosphere all the way round the globe until they get to where they're going. (And don't think some-one upstairs didn't plan *that!*)

CHAPTER FOUR.

ELECTROMAGNETIC WAVES.

When Einstein was a boy his father showed him a compass.

A compass is just a bar magnet that is hung from its center and free to swing. It will always line up pointing north and south.

Magnets occur in nature. Chinese children played with them in ancient times. They're known as Lodestones, or Magnetite, and are an oxide of iron. The ancient Greeks discovered them in Asia Minor, in the western part of what is now modern day Turkey, a place called Magnesia (also known as Manissa) to the north-east of Izmir (back then called Smyrna). Which is how they came to be called magnets. "*The iron will move to seek the stone's embrace*," wrote the Roman Lucretius some years later, around the time of Julius Caesar.

Today new alloys have been developed from which can be made magnets of far greater strength.

Magnetism is caused by more of an atom's electrons whizzing around the nucleus in one direction (or sense*) than in the other.

Einstein wondered what mysterious unseen force could make the compass needle move. He suspected it might be like light and not need any medium for it to travel through. It was also clear that this magnetic influence was running along the surface of the earth in a roughly north-south direction.

Today we know that some animals use the earth's magnetic field to tell them in which direction to travel when migrating in response to seasonal changes in climate.

We also know that since the first measurements were made, around the year 1680, the strength of the earth's magnetic field has fallen by about 15%. At this rate, in about two thousand years it will have disappeared altogether, with perhaps disastrous results.

The earth's magnetic field is generated by the spinning molten metal at the earth's core. Whether or not the present decline in the strength of the magnetic field will continue is not known. The survival of life on earth may well depend upon the outcome.

* See second footnote on page 13.

It is the earth's magnetic field that protects us from that harmful radiation known as the solar wind. At the poles, where the magnetic field terminates, this radiation is almost unopposed and appears (in the northern hemisphere) as the *aurora borealis* or northern lights. (In the southern hemisphere it's called the *aurora australis*.)

Without the protection of a magnetic field earth's atmosphere would be destroyed by this solar wind, a barrage of ionized particles streaming at us from the sun. The earth could turn into an arid, airless wasteland like Mars or the moon. Even with a magnetic field our atmosphere is slowly being lost. However, the process is so gradual it will take billions of years before it becomes a problem, and by that time evolution's greatest triumph, homo so-called sapiens, will have long since gone the way of the powerful lizards (dinosaurs).

Meanwhile an Italian count named Alessandro Volta found he could produce tiny amounts of electricity by piling alternate discs of dissimilar metals on top of one another, the discs separated by pieces of cloth soaked in brine, and connecting together alternate discs so that a potential difference (or *volta*-ge) existed between the two groups of discs. This became known as Volta's Pile.

Soon it was found that a better way to make electricity was to put bars of two different metals into a jar of sulfuric acid (copper and zinc will work). With the ends sticking out above the acid, if you connect the bars with a wire electricity will run through the wire.

Jars could be connected in groups, called *batteries.* Connecting the jars in *series* produced a higher voltage. (Think of voltage as the *pressure* behind the electricity.) Connecting the jars in *parallel* will produce more current. (Current is the *amount* of electricity that is flowing. How much of it there is.) People started making batteries.*

* The word battery really means a group of things together, such as a battery of guns. (New York's famed Battery Park was once the site of multiple guns.) Or, in this case, a battery of primary cells.

Then someone found that if you bring a compass needle near a wire when electricity is passing through the wire then the needle will no longer point north and south but will become erratic.

The force that deflects a compass needle could now be produced by using electricity. This influence of an unknown force was the phenomenon that was later to puzzle Einstein as a child. Here was something that could make things move though you couldn't see it!

Electricity could be used to make a temporary magnet. If you take a bar of iron (one that's not already magnetized) and wind an insulated wire round and round the bar, when you pass an electric current through the wire then the bar (virtually) instantly becomes a magnet.

When you switch off the current the bar is no longer a magnet, except for a little bit of what's called residual magnetism, where the molecules haven't quite gone back to where they were before.

Scientists called these invisible magnetic phenomena lines of force, or flux. James Maxwell coined the name magnetic field.

Soon people were sending messages over wires by energizing temporary magnets in a series of dots and dashes, causing the pony express to experience a sudden decline in the volume of its business. It was the start of the new age of electronic communication, which was to change the world so dramatically (and not necessarily for the better) in the years ahead.

About this time a young Englishman by the name of Michael Faraday was discovering that not only could you create magnetism using electricity, but you could create electricity using magnetism.

What you do is take a magnet that's bent around in the shape of a horseshoe, so that the ends are close together and the magnetic field between them very strong.

Then you take a piece of wire and pass it sideways through the magnetic field very quickly. As the piece of wire is cutting the lines of force a voltage is generated in the wire. People stopped making batteries and started making generators.

This is the source of almost all the electricity generated today. (Solar panels would be an obvious exception, as would the primary cells used in, for example, flashlights or television remote controls.)

From modern nuclear power plants, to the fossil-fuel plants that pollute the atmosphere, to the hydroelectric plants which could have unknown effects on the environment, to the wind-driven generators that are now starting to spring up everywhere, they all make use of Faraday's discovery of what he called electromagnetic induction.

And let's get back to those radio waves we talked about back on page 24. Because radio waves are also electromagnetic waves.

As we described on the previous page, when an electric current passes along a wire a magnetic field is set up around the wire. This is what affects a compass needle. If we switch the electricity off then the magnetic field collapses.

What if we keep switching the current on and off quickly? Or hook it up to our household electricity supply, which is alternating back and forth rapidly? The magnetic field will keep building up then collapsing back down onto the wire again. Is this a problem?

When our household electricity supply alternates back and forth it does so in *cycles*. What do we mean by a cycle?

Well at first no current is flowing. Then it starts, and increases, in the form of what's called a sine-wave, until it reaches a peak. Then it drops back to zero and no current is flowing. Next the current starts to flow again, but in the opposite direction now, reaching a peak as before then falling again to zero. This is one full cycle.*

Some places, notably the United States, use 60 cycles per second while others, such as Europe, use 50.**

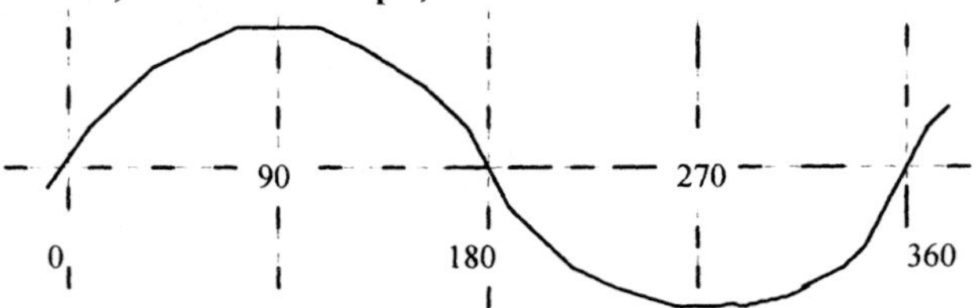

Sine wave, showing degrees of rotation of generator shaft.

There are advantages to using a higher frequency. For example electrical equipment can usually be smaller for the same power. But it would be difficult for the Europeans to change now, as that would mean having to scrap practically all their existing equipment.

* See appendix (7) for more on sines and right-angle triangles.

** In the early days, both in Europe and the United States, 25 cycles was common. Realizing this was too low the Europeans decided to double the frequency, while in the U.S. we chose to go even higher.

Incidentally the development of alternating current was bitterly opposed by Thomas Edison who favored the use of direct current. His rival, Westinghouse, staked everything on alternating current, and for once in his life Edison turned out to be dead wrong.

So if the electric current is switching back and forth rapidly what does this do to the magnetic field?

As the electric current starts to flow the magnetic field spreads out from the wire until it stabilizes. Then when the current is turned off the field collapses back down onto the wire again.

Now if we alternate the current back and forth at 60 cycles per second there's no problem! Not at *60* cycles per second. But what if we did it at *30 million* cycles per second?

To have electricity switching back and forth at 30 million cycles per second (we call it 30 *mega*-cycles) is pretty fast, isn't it?

In fact it's a very modest frequency. We call it the ten meter band because the distance covered in a 30-millionth part of a second, when traveling at the speed of light, is about ten meters.

Although the ten meter band is in the high frequency range this is only because it was given that name a long time ago. Back then it *was* a high frequency. Today most radio work is in the VHF, or *very* high frequency range, which is much higher, or in the UHF or *ultra* high frequency range, which is higher still. Radio wavelengths are now often so small they're measured in centimeters.

One cycle means that the magnetic field spreads out from the wire, then collapses back down again, *twice* (once in one direction and once in the other direction). So at 30 megacycles the magnetic field spreads out, then collapses, *60* million times per second.

How far out from the wire does the magnetic field spread? Well that depends on how much current is flowing. But let's say it spreads out ten or twenty yards. And there's a limit to how fast a magnetic field can move! (More on this later.) It travels fast, *but not instantly.* What happens is this.

As the current starts to flow through the wire the magnetic field is established around it. Then as the flow of current is reduced the magnetic field collapses back down onto the wire.

But it can't get there! At least not all of it. Because (as we just said) there's a limit to how fast a magnetic field can move. And if the current is alternating very rapidly, then before the collapsing field has time to get all the way back to the wire *it meets the next wave which is already spreading out from the wire to meet it.*

The next wave won't let the first wave back onto the wire, and pushes it off into space where it can be detected by an antenna tuned to that wavelength. Your TV antenna for example (in the days before cable or satellite). A properly tuned antenna is half the length of the wave it's meant to detect. We say the wave is "forced off the wire."

Incidentally when you talk on a mobile 'phone electromagnetic waves are being forced off the wire just inches from your brain, yet no-one seems to be concerned about what the harmful effects this may be over a long period.

Picture of man zapping his brain with very high frequency radio (electromagnetic) waves.

It is strange that people should be so fearful of nuclear power when, due to this radio wave propagation, the regular use of mobile 'phones is probably more harmful than the most radiation anyone received in the much publicized Three Mile Island nuclear incident.

Unlike the Chernobyl disaster in the former Soviet Union, in a well designed plant, where there are enormous suppression pools to serve as heat sinks, where multiple emergency backup cooling water systems are provided, and where the reactor pressure vessels themselves are enclosed within massive reinforced concrete containment buildings, an event such as that which occurred at the Three Mile Island plant in March of 1979 is virtually a worst case scenario.

Even though the risks involved with nuclear power are so small, people evidently prefer to discharge *tens of millions of tons* of greenhouse gases into our thin film of atmosphere *every day** burning fossil fuels, causing major climate changes, quite apart from wasting the planet's precious and irreplaceable mineral resources. For hydro carbons are essential to the chemical and plastics industries, as well as for fertilizers. In fact it's only by the use of such fertilizers that we're able to feed the six and a half *billion* people there are in the world today. At least in the short term, nuclear power is going to be indispensable if our whole house of cards is not to collapse!

* One authoritative Australian source puts the CO_2 emission at 7·85 *billion* tons per year, a 15% increase in five years. Other possibly less reliable sources have put it as high as almost five times that amount.

While certain harmful emissions, such as oxides of sulfur and nitrogen, can be *reduced*, though probably not eliminated altogether, the same is *not* true of carbon dioxide.

Carbon-dioxide is the *inevitable and inescapable* bye product of producing heat by the efficient* combining of carbon with oxygen. So far we've found no large scale way of capturing and disposing of it. Since the start of the Industrial Revolution the concentration of CO_2 in the atmosphere has more than doubled, and is forecast to double again by about the year 2075. For while much of the carbon dioxide produced is fortunately being absorbed by places like the Amazon rain forest, such places all over the world are being rapidly and systematically destroyed.

Remember (on page 26) we talked about gamma rays, and said they were the deadliest emissions known to exist. Well on July 2nd, 1967, while searching the sky for signs of nuclear tests being carried out by potential adversaries, a massive burst of these gamma rays was detected in the southern hemisphere. The burst was so powerful it had to be coming from within our own galaxy, or so it was thought.

But then others were detected, some of them found to have come from so far away that, traveling at the speed of light it had taken them more than *ten-billion* years to get here.**

To be detectable at this immense distance the source of energy would have to be so powerful that, had it in fact been located in our own galaxy, the earth and all the planets would have been vaporized by the heat. Such enormous energy was beyond belief.

However, this is not the whole story! Such violent bursts of energy can only come from one source, the death of a star. Then the amount of energy available *depends on the size of the star* according to the equation $E=mc^2$ (more on this later).

* Combining carbon and oxygen *less* efficiently, as in car engines, for example, can produce *some* CO, or Carbon *Mon*-oxide, a process yielding only about one third as much heat as combining to CO_2.

** New bursts are now being detected *every day*! The second most powerful so far occurred on February 22nd 2001, detected by the VLA (a Very Large Array of seventeen radio telescopes) in New Mexico.

The problem is that stars cannot exist beyond a certain size, and to produce as much energy as these gamma rays the exploding star would have to be *bigger than it's physically possible for a star to be!*

This challenged the validity of the relationship between mass and energy that had been established by special relativity back in 1905. More of these emissions being subsequently discovered they were given their own acronym, GRBs, for Gamma Ray Bursts.

So now we have a problem! The intense bursts of gamma rays could *only* come from a dying star (a "Gamma Ray Burster") yet no star could ever be big enough to produce *that* amount of energy!

On pages 111 thru 115 we'll be talking about novae and super-novae, exploding stars which emit enormous amounts of energy.

There are two kinds of supernovae, the Type 1 and the Type 2. Type 1 are produced by big stars, and Type 2 by *even bigger* stars.

After a Type 1 supernova explosion dies down there's usually not much of it left, it all having exploded out into space. But in a Type 2 explosion *if the originating star was big enough* there could be a massively dense core left, what's known as a neutron star. And if the star's *even bigger*, then what is left afterwards could be something even stranger, a (stellar) black hole.* The name has to do with what's called its *escape velocity.*

All bodies exert gravity on all other bodies. If you fire a pistol into the air the bullet will fall back to earth. The bullet is attracted to the earth (and vice-versa). If you used a high power rifle the bullet would go higher, but it would still fall back to earth eventually. Gravity is the cause.** (We all know that!)

But gravity diminishes the farther away from the body you get. So if you could fire a rifle bullet (or anything else) *fast enough* it would just keep going. It would *never* fall back to earth.

How fast would you have to fire a thing into the air for it to escape from earth's gravity altogether?

* There are also much larger *super-massive* black holes (see next page) at the centers of (probably all) galaxies, including our own galaxy the Milky Way. And super massive black holes themselves can vary in size greatly.

** Incidentally gravity on the moon is only a sixth that of earth, so it's a hundred times cheaper to blast payloads into space from there. For this reason the moon could serve as a platform for the assembly and launching of future space craft bound for more distant bodies.

To escape from earth's gravity altogether a body would need to be traveling at about 11·2 kilometers (6·9 miles) *per second.* This is within the capability of modern rocket technology. (After all, we've sent astronauts to the moon!) It is the earth's *escape velocity.*

But what if we were on a larger planet, like Jupiter, say? Gravity on Jupiter is more than two and a half times that on earth.* We'd need to make the bullet, or whatever it was, go much faster before it could escape gravity and fly off into space. The more massive a body is (and the more dense) the greater the escape velocity!

A neutron star has a mass greater than that of our sun yet is squeezed into a space only a few miles across. As the giant star turns into a supernova the atoms break apart. The enormous gravity causes the positively charged protons and negatively charged electrons to *combine.* With their charges canceling each other out they turn into neutrons, which have no charge. Hence the name *neutron* stars.

A (stellar) black hole has an *even greater* mass than a neutron star and is squeezed into an *even smaller* space. Its gravity is so great as to tear through the fabric of space rather than merely distort it. (It's said that if the earth were the density of a black hole it would be the size of a sugar cube.) There are also super-massive black holes at the centers of galaxies (see previous page). The one at the center of our galaxy is thought to be between *three million and four million* times the mass of our sun, yet not much larger in size.

A black hole's escape velocity is greater even than the speed of light! And if light can't escape from a black hole it can't be seen! Which is why it's called black.

This is the difference between a neutron star and a black hole. A black hole results from the explosion of a star *bigger even* than is needed to produce a neutron star. (Another possibility is that a black hole may result from the combining of two neutron stars.)

A neutron star can be detected by the radiation it emits, while a black hole emits *nothing!***

* You'd think it would be even more since Jupiter is *three hundred times* the earth's mass. But it occupies a much greater *volume* so that its exterior (it doesn't have a *surface*) is very far from its center.

** This may not be quite true. There's now thought to be something called Hawking Radiation, named for the famous British astronomer Stephen Hawking. But it's not something with which we need to be concerning ourselves here.

Now back to the GRBs that seemed to be coming from the death throes of a star bigger than it's possible for a star to be! (GRBs are so powerful they make a supernova look like a campfire.)

Well astronomers were already aware that during the birth of a black hole intense bursts of radiation are emitted, *but not in all directions as one would expect*! The intense magnetic field of the giant collapsing star causes the radiation to be focused *into two narrow beams*, like two searchlights back-to-back and sending out concentrated shafts of light in opposite directions. (Like the flashing pulsars we've all heard about.) If this were the case with the GRBs then the *total* amount of energy wouldn't need to be *anywhere near* as great as if it were being radiated out *in all directions* at this same intensity. The only reason the rays were so powerful was because all of the energy was concentrated in just those two narrow beams!

So now it was realized that, in order to emit the GRBs, the dying star didn't have to be impossibly large after all! What we had was just a catastrophic explosion, resulting in turn from the death of a *very big* star! But a star of a size that was at least *possible.*

And if all the energy was concentrated in just two narrow beams, then what was being measured, instead of being just a tiny fraction of the total radiation, could in fact be a major part of it. To everyone's relief Einstein's equation of $E=mc^2$ was vindicated.

The radiation that originally caused all the commotion is, as we just said above, far greater than a novae or supernovae though, and some scientists are calling this source of energy a *hyper*-nova! It represents the death of a giant star, and the birth of a black-hole.

Should such an event occur within our own galaxy, even though we now know it's not near as powerful as we'd thought, it could still be a problem. Perhaps even a disaster! Fortunately the likelihood of this happening is thought to be remote. (It's pretty certain there's not been an event of such violence in our Milky Way galaxy since the solar system came into existence four and a half billion years ago!)

And finally we should mention that cosmic rays (see page 26) are not all bad! For they are a cause of occasional random genetic mutations that occur in living organisms, and which make possible the process of natural selection by which we have all evolved.

But we've digressed enough! Now we need to get back to things more directly related to special relativity.

In the next chapter we're going to explore those all-important things called *energy* and *mass*!

CHAPTER FIVE.

ENERGY AND MASS.

For centuries it was thought that energy and mass were two different things. It was true that if a thing had more *mass* then it took more *energy* to get it moving. But they were not *directly* related.

First there was the law of conservation of energy, which meant that *the overall quantity of energy must always be the same.*

Energy can take different forms. It can be in the form of heat. Or it can be *kinetic* energy, the energy a body has due to its motion. It can be *potential* energy, the energy you give a body when you lift it up. Then there's *strain* energy, like the energy in the spring of an old fashioned watch after you've wound it up. Energy can take different forms, but whatever the form *it didn't change the total amount.* That was the old law of conservation of energy.

Then we had the law of conservation of mass. Unfortunately we use the same units for mass as we do for weight, while they're not the same thing. You'll find more about this in appendix (8).

Imagine a block of stone with a *mass* of one ton. The block is attracted by the earth's gravity with a force (its *weight*) of one ton. If you placed it on a cart and tried to accelerate it along a dead level surface, even if there were no friction you'd still have to push it quite hard. If it had been a feather you'd have been able to move it with your little finger.

Now if you took the block of stone up in a space-ship, in what's known as micro-gravity, then it would weigh next to nothing.

Out in space, beyond the pull of earth's gravity, and beyond the influence of any other celestial bodies, then the block of stone would weigh the same as a feather, or virtually nothing. But the *mass* of the stone would be the same as it was back on earth. *You'd still have to push it just as hard if you wanted to get it moving.* The stone's inertia, its resistance to being moved, is unaffected by gravity, or by the lack thereof.

The feather, though, could still be moved with your little finger, because its *mass* is small. *It is mass, not weight, that matters when you're trying to accelerate things.*

Are you saying weight doesn't matter?

For a long time people believed in the law of conservation of mass, which said that the total amount of matter (or mass) was fixed. You could burn a piece of wood, say, and the mass of the ash that remained would be less than the original mass of the wood before it was burned. *But the original mass had not been reduced,* part of it had simply changed into different forms.* For example much of the carbon in the wood's original mass would have combined with oxygen from the air to form carbon dioxide and floated away (to mix with all the rest of the stuff we're dumping into the atmosphere).

There used to be an old dictum that said "*matter can neither be created nor destroyed!*" People believed this for centuries.

Then in 1905 special relativity revealed that this old law was not strictly correct, and that the total amount of mass can in fact change. For it has been found that under certain conditions it can be changed into energy, such as in atom bombs or nuclear power plants. But we don't want to get ahead of ourselves, so we're going to leave further discussion of this until later.

Before leaving the subject of *mass* there's something we need to familiarize ourselves with. It is Newton's law that says *force equals mass times acceleration.* This is written as F = ma, and is important enough that it is explained in detail in appendix (11).

Now let's get back to what the speed of light is *relative to!*

* Or at least any change in mass would be virtually infinitesimal, and certainly not measurable.

CHAPTER SIX.

WHAT THE SPEED OF LIGHT IS RELATIVE TO.

Light can take the form of a wave. But when traveling through space we've found out that it does so in the form of photo-electrons, or "photons." While photons* are not exactly *particles* (they have no mass, for one thing) they *behave* like particles.

Now in the study of special relativity we're concerned not so much with *how* light travels but with its speed! Especially we need to find out what the speed of light is *relative to.*

A bullet is like a particle. If you're on a moving train and fire a rifle in the direction the train is going then, relative to the ground, the bullet will be traveling faster than it would have done had the train not been moving. You have to *add* the two velocities!

Similarly if a man out walking his dog in East Los Angeles was the victim of a drive by shooting he'd be better off if the gunman's car was traveling *away* from him as the shot was fired. Then the velocity of the bullet (through the air) would be reduced by the speed of the car.

What if we were measuring the speed of *light*? If light travels through space in the form of photons (which *behave* like particles) then shouldn't its speed *also* depend on whether the *source* of the light was moving *away* from the observer or moving *towards* him?

Suppose we measured the speed of light coming from a distant galaxy. We know that distant galaxies are rushing away from us at high speed, so we'd expect to find the speed of *this* light was *less* than usual because the *source* of the light was moving *away* from us.

But this is not so. When we measure the speed of light coming from something that's moving away from us at high speed, we find that the speed of *this* light is *the same* as that of the light coming from bodies that *not moving away from us at all!*

What about the motion of the *receiving* body? In our case the earth. The earth is moving around the sun (traveling at a speed of about twenty miles per second). What if we measured the speed of light from a *star in whose direction we just happened to be moving?* Then, six months later, having gone half way round the sun and now headed *away* from that same star's direction, we should be able to detect a difference in the speed of the light coming from it.

* From the Greek *photo* meaning light.

But again this is not so. *Whatever we do doesn't seem to make any difference to the speed at which light is observed to travel.*

So just what *is* the speed of light relative to? Is there a point someplace that light moves *relative to* but that *itself* doesn't move? Galileo didn't think so, but then Galileo could have been wrong! (It's possible. In later life Einstein was wrong about a lot of things!) So let's see if there *is* a *fixed* point anywhere in the universe.

If you ever get to London you'll want to visit Nelson's Column, a monument to England's greatest naval hero located in Trafalgar Square. Nelson was a British vice-admiral who was killed aboard his flagship, the HMS Victory, while methodically annihilating the combined French and Spanish fleets off the southern coast of Spain on the afternoon of Monday, October 21st 1805. So decisive was the outcome of the action that British naval supremacy was not to be seriously challenged again until the Battle of Jutland, one hundred and eleven years later.

Today HMS Victory is in dry-dock in Portsmouth, England. On the deck is a brass plaque marking the spot where Nelson fell. (*"No wonder he fell,"* complained one visitor, *"I nearly tripped over the thing myself."*)

Above photograph by permission of Devon and Clive Kelsey.

But wait, you say, the Victory is in *Portsmouth*, and Nelson wasn't killed in *Portsmouth*, he was killed off the coast of *Spain!* So how could a plaque *in Portsmouth* mark the spot where Nelson stopped a musket ball if it really happened *off the coast of Spain*?

No, the plaque marks the spot where Nelson fell *relative to the rest of the ship.* To find the true spot where it happened you'd have to take the ship from Portsmouth and sail her all the way to Spain. That way you'd be in the same exact spot where Nelson was when he got killed. The same fixed spot. At least *relative to the earth!*

But to do it right you'd have to make sure the *earth* was in the same place it was when the battle took place. You'd have to go on *the same day of the year*, when the earth's position *relative to the sun* was the same as it was on the day that Lord Nelson was killed.

Only that wouldn't be enough either, because the sun is moving, within its local group of stars, at a speed of about 13 miles a second. And this local star system *itself* is moving within its galaxy, the Milky Way, at the considerable speed of around 200 miles per second. Even that's not the end of it, because the entire Milky Way is drifting (with respect to the more remote galaxies) at a speed of about 100 miles per second. And all these motions are in different directions! It seems Galileo was right! *Nothing is fixed.*

So just where *was* the diminutive admiral when he cashed in his chips around 4 pm on the afternoon of Monday, October 21st, 1805? Shouldn't there be a way to identify the spot by reference to some *fixed* point?

If the universe had a center, a fixed point that everything moves *relative to*, but that *itself* doesn't move, then presumably it would be to this fixed point that the speed of light, too, must be relative! (It would solve a lot of problems.) So let's see if it's possible that the universe has a center!

What is the universe? To paraphrase the late Carl Sagan it is *"all there is, all there ever was, and all there ever will be!"*

The universe began around twelve or fifteen *thousand-million* years ago (current estimate $13{\cdot}7 \times 10^9$) and started in what's known as *the big bang,* a derisive name coined by the English astronomer Fred Hoyle, who claimed the big-bang never happened, and said the reason the universe was expanding was because new matter was being created all the time. No-one believed this for a moment, and Fred was duly knighted by the Queen the following Tuesday, along with several long-haired rock musicians and other important people.

Before the big bang there was *nothing*. Then there appeared a tiny point, smaller than an atom, yet containing all the raw material for everything that now exists. What would later become the stars and galaxies, the planet earth and all the things we know, everything required to make the universe the way it is today, was all contained within that *infinitesimal* point astronomers call a singularity.

A singularity is a point of infinitesimal size, where all the laws of physics break down and where time and space are extinguished!

At the big bang everything started to fly apart. Soon the energy of the big bang began to change into elementary particles, and then into atoms of hydrogen and helium. Under gravity the gases became concentrated, becoming so dense that nuclear fusion of the hydrogen began to take place and the first stars began to shine. "*And God divided the light from the darkness,*" the Bible says (page 23).

The infant universe continued to fly apart, and is still doing so today (though not as fast as it was at first). Now, after thousands of millions of years, things have got very far apart indeed.

Now imagine you're watching fireworks on the fourth of July. Suddenly you are startled by a bursting rocket that scatters brilliant debris in all directions. There is a point in the sky where the rocket exploded. Even if the location of this point was not noted at the time, it could still be identified later by analyzing the motion of the debris from the explosion. The height, latitude and longitude of the point where the explosion took place could in this way be established. (At least its location could be established *relative to the earth!*)

Similarly, if the big bang really did take place then it had to have taken place *somewhere*. There had to be a fixed point, *the center of the universe,* and we should be able to identify it by tracing the paths of all the celestial bodies backwards to the point from which they must have all started out! But unfortunately this raises a problem.*

We live in a world of three dimensions. There is forward and back, there is left and right, and there is up and down. These are the measures in terms of which everything that we've ever experienced exists. On earth there *are* only three dimensions!

* Space itself being contained within the singularity at the instant of the big-bang, that means the big-bang happened not in any one place but *everywhere!* Which is why today faint residual "noise," associated with developments occurring round about the time of the big-bang, can be detected coming at us *from all directions at once*.

But while we may live on a three dimensional planet, we're part of a universe that, everyone agrees, *has to* have more than just three dimensions.* In contemplating additional dimensions we no longer rely upon our senses but must use logic, an attribute particularly well developed in humans, fortunately.

It's *spatial* dimensions we're going to be talking about here, not *time* (see page 132). By spatial dimensions we mean dimensions in space. And only a *fourth* dimension. We're not going worry about all the rest of them there's supposed to be. (You'll be relieved to know this book is not concerned with any of *them*.)

No-one can understand a fourth dimension in space. Not even the cleverest man (in English that can mean *person*) who ever lived. But we can play a little game that may help us to visualize it.

Imagine there's a village, rather like our world, except that it's a *flat* village. There are only *two* dimensions. Here everything is flat, including the inhabitants. They can understand left and right, and they can understand forward and backward. *But these flat people have no understanding of up and down.*

Even so they're a clever people and they've discovered there are other villages like theirs, other two-dimensional villages.

These other villages are very far away, and getting farther and farther away all the time. So up to now the flat people haven't found a way to visit any of these other villages. But they know they exist, and can study them through their flat telescopes.

Now these flat people reason that if all the villages are moving away from one another, and if at one time they were closer together, then there had to be some original point from which they all started out. Their (two-dimensional) universe must have a *center*, they say!

But there's something these flat people don't know. Because there is actually a *third* dimension, and even though they *think* their system of villages is spread out over a *flat* surface, they are actually on the surface of a giant balloon, and someone (God) is blowing up this giant balloon, and has been doing so ever since he created their universe billions of years ago when it was smaller than an atom.

The surface of the balloon is smooth *all over*. Let's say it's being blown up from the inside somehow. (God, of course, would be able to do that.)

* There may be as many as ten or eleven dimensions (see page 132). But we don't want to make this any worse than it already is!

Because the balloon is being blown up, and since the villages are spread out over *the surface* of the balloon, they're getting farther and farther away from one another all the time.

The balloon is so big the people think the land on which they live is flat, while it's actually the surface of a giant balloon and so is very slightly curved, though not enough for them to be able to detect. (The light in their universe is also curved and follows the surface of the balloon. Otherwise they wouldn't be able to see the other villages.)

As the flat people have evolved knowing only *two* dimensions, *so can understand only the surface of the balloon*, is there a (fixed) center to this system of two dimensional villages spread out over the ever growing *surface* of the balloon?

The answer is no. It's not possible to take a point *on the surface* of the balloon and say that it's the center. (Remember, the *surface* of the balloon is all these flat people know.)

Now think of how we humans understand only three dimensions when logic tells us there has to be at least one more than that, and you'll see we're not all that different from the flat people. We're subject to the same limitations they are. *There's another dimension (in this case a fourth dimension) that we're unable to visualize!*

So now we've tried every way we can think of to find out what the speed of light is *relative to*, yet we're no nearer than when we started. But the answer has been staring us in the face all the time!

Elementary, my dear Watson! *

The famous fictional detective Sherlock Holmes is supposed to have said, "*When you have eliminated everything else, whatever remains, however improbable, is the truth!*"**

* Contrary to popular belief, Sherlock Holmes never actually said "*Elementary my dear Watson.*" At least not in any of the original Sir Arthur Conan Doyle (1859-1930) stories.

** When Conan Doyle was a medical student this was a principle impressed upon them by Dr. Joseph Bell, a faculty member on whom the character of Sherlock Holmes was later to be modeled.

That's why we couldn't see it, it was so *improbable*. We'd been unable to find *anything* that the speed of light was relative to, but we were overlooking the obvious! The speed of light *had to be* relative to *something* or we wouldn't have been able to measure it!

The very fact that we were able to measure it, *and always get the same reading* (a fact that had been known for some time) could mean only one thing. *The speed of light wasn't relative to anything except to the person looking at it!*

When Einstein realized the implications of this he nearly had a heart attack.* (It caused him the greatest intellectual difficulties, was the way he put it.)

So there you have it! The speed of light is relative *only* to the observer, and is *totally* unaffected by anything else. It's what special relativity is all about. It's as simple as that!

Now you know all there is to know about special relativity. It's not hard to understand, we just don't know *why* it should be this way.

Of course you might say what does all this have to do with anything? Well to most of us in our everyday lives it's of no importance. But when observing bodies traveling at very high speeds it matters a great deal. And this is what we're going to be doing in part two.

As to exactly *why* the speed of light is the way it is (see above) well once it was believed that sooner or later everything would be understood. All we had to do was study and experiment, and apply the world's greatest brains, for a long enough period of time.

Yet today we are less sure. There are some things which seem to be beyond the ability of man to comprehend. There are cases where we already have all the knowledge we're ever going have, where we've applied the finest brains, over long periods of time, and yet are still completely baffled.

One such case is the famous puzzle known to mathematicians as Goldbach's Conjecture. We don't want to get away from the subject of special relativity, so if you want to find out more about this piece of mathematical mischief you can go to page 135. We only mention it here because the supremacy of the speed of light (and every other form of electromagnetic radiation) is another thing the reason for which we have no idea, and probably never will have.

* This wasn't any sudden revelation! James Clerk Maxwell (1831-1879) and Ernst Mach (1838-1916) were among others whose work had contributed to this awareness.

In passing it should be mentioned that, while special relativity says the speed of light is the fastest anything can go in the universe, in quantum mechanics it is claimed this not always the case.

Sometimes subatomic particles are said to disappear from one orbit, and *simultaneously* reappear in another orbit, either taking in, or giving out, a quantum* of energy in the process, *without having traveled from the first orbit to the second orbit.* This is the so called quantum leap. (You'll have heard people talking about a quantum leap, and evidently thinking it means a *very big* leap, but that's not really what it means at all.)

Also it is said that changes in the polarity of certain subatomic particles *can* cause instant reversals in the polarity of *other* particles located at such a distance that any communication between them must have been at a speed faster than the speed of light.

Still, that's in quantum mechanics and here we're talking about relativity. The problem is that quantum mechanics and relativity are not compatible. Meaning they can't *both* be right! (Something no-one's been able to quite figure out yet.)**

So this brings us to the end of the first part of our book, the part about how we got to the idea of relativity.

In the second part of the book we're going to be looking at the effects that special relativity has on things we've always taken for granted.

And some of these effects defy common sense.

* Latin for "*how much.*". (In this case very little.)

** Much scientific research these days is being directed towards the question of why natural laws that apply to the very large, do *not* seem to apply to the very small. However, this is not a matter with which we need to concern ourselves here.

Part Two.

Special Relativity produces some unlikely results!

FOREWORD TO PART TWO.

In part two you'll notice that in order to keep things simple we'll sometimes say that a body is *moving* or *standing still.* We haven't forgotten what we said back on pages 16 & 17, what we mean is *relative to the observer!* And in part two bodies won't always be moving steadily and in a straight line like they were in part one. Sometimes they'll be *accelerating.*

To engineers words have a very precise meaning. Take the word *work*, for example. *Work* means the expenditure of *energy*. You do *work* when you pick up a brick off the ground and place it on top of a box. If the brick weighs ten pounds, and the top of the box is five feet above the ground, then in lifting up the brick you'll have done fifty *foot-pounds* of work. The *work* went to increasing the *potential energy* of the brick. It now has the *potential* for doing *work* (which you'd find out if it dropped on your foot). In returning to its original position the brick will, in one way or another, return the *energy* it received when it was lifted up. It will perform *work.*

Other types of work/energy are directly related. There is heat, usually measured in BTUs. One BTU equals 778 *foot-pounds.**

Then there's electricity. We buy *work* in the form of kilowatt-hours delivered to us over wires by the electric utility company. A kilowatt-hour is about 3,412 BTUs or some 2·65 million foot-pounds.

Now let's talk about *distance, speed* and *acceleration.*

(1) By <u>distance</u> we usually mean *how far you've traveled.* For our purposes here let's use feet. O.K., remember, *we're going to be measuring distance in feet!*

(2) <u>Speed</u> is the rate at which distance traveled is increasing, as in feet per second. Remember, here we're going to measure speed not in feet but in *feet per second.* On the highway of course we use miles per hour. (60 miles per hour is exactly 88 feet per second.)

As *speed* causes *distance* to change with the passage of *time*, we call speed the "first differential" of distance with respect to time. In calculus we write (first) differentials as dy/dx, where in this case y is distance and x is time.

* A *foot-pound* is a unit of *work.* To avoid confusion engineers use the term *pound-foot* as a measure of *torque* because that's something different. This is another example of the importance of utilizing a very precise term. (See second footnote on page 13.)

Don't worry, we won't be using any calculus in this book. But it's not something to be afraid of. Calculus (the differential kind) is just a tool we use to help us get a handle on *the rate at which something is changing!* Sir Isaac Newton devised calculus around the year 1665 when he would have been only about twenty-two years old. A German mathematician, Gottfried Leibniz (1646-1716) seems to have done so independently, though evidently some years later.

Leibniz was the first to publish, Newton being notoriously slow to ever publish *anything*, so for a time it was unclear as to whom we were indebted for this important advance in mathematics. (Actually in this case it took Newton almost forty years to publish.)

(3) Then we have *acceleration*. This is *the rate* at which speed is changing. And remember *speed itself* is just an expression of the rate at which *something else* (namely *distance*) is changing.

So in calculus we say acceleration is the *second* differential of distance with respect to time. Second differentials are written as d_2y/dx^2, again where y (in this case) is distance, and x is time.

Don't say you don't understand acceleration. Everyone knows that if you jump off the top of a high building, the speed at which you'll hit the ground will depend on just *how* high the building is.

In actual fact, if you did do such a foolish thing, then after one second you'd be traveling at a speed of 32·2 *feet per second.* Were it not for air resistance your speed would continue to increase at a rate of 32·2 feet per secon*d every second.* Until you reached the ground, of course, whereupon your precipitous decent would be rapidly slowed by an electromagnetic force *billions* of times stronger than gravity (see page 104).

So remember, if *speed* is the rate at which *distance* is changing with respect to *time* (the *first* differential) and is measured in *feet per second*, then *acceleration*, being the rate at which *speed itself* is changing (or the *second* differential) must be measured in *feet per second per second.* Or feet per second *squared!*

Acceleration (unlike *speed*) only results from the application of a *force!* Like Newton's apple, our bodies are attracted to the earth with a *force* due to *gravity*. Should we find ourselves in mid-air, then this *force*, acting on our bodies, will cause us to be *accelerated* downwards. You'll find more on this subject in appendix (11). But remember, acceleration is always *distance* divided by *time* squared!

Or $a = d/t^2$ (We shall be using this later on page 86.)

CHAPTER SEVEN.

THE SPECIAL RELATIVITY OF LIGHT.

Special relativity doesn't affect us in our everyday lives. It's only astronomers and physicists and people like that who have to deal with it. For the rest of us Newton's laws work just fine, just like they have for the last three hundred years or so.

But special relativity has shown that at extremely high speeds Newton's laws are no longer accurate.

What led to this was the discovery that light always travels at the same speed. No matter what the motion of the source of the light, or of the person observing it, the speed at which light is observed to be traveling never varies! *Under no circumstances can the speed of light ever be anything other than (about) 186,000 miles per second.*

Do you remember when we said the speed of a body depended on what it was being measured *relative to*? That was before we found out about the speed of light.

There was the man in the car going 50 miles per hour, and the man in the car behind him going only 40 miles per hour. The man in the first car was going 50 miles per hour relative to a man sitting by the side of the road. But relative to the man in the second car he was going only 10 miles per hour

And there was the flight attendant aboard an airplane which was traveling at 500 miles per hour. She was walking towards the back of the 'plane at 4 miles per hour to serve a drink to a passenger. *Relative to the pilot and the passengers who were seated* the flight attendant was moving at 4 miles per hour, but *relative to the man on the ground* she was moving at 496 miles per hour.

Then there was the man in the supersonic jet trying to call the flight attendant. The sound of his voice was traveling towards the flight attendant (farther forward in the cabin) at 760 miles per hour *through the air in the cabin.* But the air in the cabin was moving (with the rest of the 'plane) at 1,350 miles per hour. So to the man on the ground (if he'd had any way of measuring them) *those particular* sound waves were traveling, *relative to him*, at 2,110 miles per hour. We said you had to *add* the speeds.

But all this falls apart when we're talking about the speed of light! Light *always* travels at 186,000 miles per second *relative to the person observing it!* You *never* add or subtract any other speed.

So then the speed of light is always the same to an observer, *any* observer, and is independent of the motion of either the source of the light, or of the observer himself. Two different observers, moving at different speeds, will *each* see a flash of light as traveling at the same speed *relative to them!* This can have strange repercussions, as we shall see in the next example.

The railroad company has built a new kind of passenger coach with automatic doors, one at each end of the coach. In the center of the coach is a light bulb which can be switched on by a railroad employee inside the coach when he wants to open the doors.

Each door is the same distance from the light bulb. Mounted on each door is a light sensor (the kind you'll find at a store entrance to alert the shopkeeper to the presence of a customer). Should either sensor detect light coming from the light bulb it will cause its respective door to open.

It is decided to test the doors with the coach moving along a straight and level stretch of track at a steady speed of about half the speed of light.

Inside the coach the railroad employee pushes the button and the light bulb flashes. He (the railroad employee *inside the coach*) sees the light traveling at 186,000 miles per second *relative to him.*

Which means the railroad employee *inside the coach* also sees the light traveling at 186,000 miles per second relative to *everything else inside the coach.* Including the doors.

As the doors are the same distance from the light bulb, and as the light is moving at the same speed towards each of them, then it will take the same length of time for the light to reach them. A person *inside the coach* will see both doors open at the same instant. In fact he'd see the same thing whether the coach was moving or not!

A person *inside* the coach will *always* see events *inside* the coach the same way. *He wouldn't even know whether the coach was moving or not except by reference to some outside point.* (See page 18 where we talk about the man and his wife on the cruise ship.)

So far everything is straight forward. It seems pretty obvious that the railroad employee *inside the coach* is going to see both doors open at the same instant.

However, there's another man involved in our little experiment. This other man is sitting on an embankment alongside the railroad track as the train is rushing past.

The man sitting on the embankment beside the track watches the train as it approaches. It is traveling at high speed. Let's say the train is heading *north!* He (the man on the embankment) sees the light bulb flash. (He can see it through the windows.)

Now what did we learn about the speed of light? We learned that light *always* travels at 186,000 miles per second *relative to the person observing it.* (This is fundamental to special relativity!)

The man sitting on the embankment is *not moving* relative to the track. So if the light beam, on its way to the relays to open the doors, is traveling at 186,000 miles per second *relative to the man on the embankment*, then from his viewpoint it must *also* be traveling at 186,000 miles per second *relative to the track!*

So once the light bulb has flashed then, *as far as the man on the embankment is concerned*, we must forget about the light bulb and just concentrate on what the actual beam of light is doing.

We said the light spreads out at 186,000 miles per second in all directions, including north and south along the track, from the point where the light bulb was at the instant it flashed.

But the train is traveling north at half the speed of light. Which means the coach, together with its doors, is traveling *north* at about 93,000 miles per second.

Let's say the flash of light has just occurred and it hasn't had time to reach either door yet. So it's somewhere between them.

The light is heading north at 186,000 miles per second towards the front door, which is also heading north but at only 93,000 miles per second. The light will catch up with the front door because it's going twice as fast. But the distance between them is closing at only *half* the speed of light. (Make a note of this!)

At the same time the light is traveling south at 186,000 miles per second. But in this case the rear door is rushing north to meet it at 93,000 miles per second. So that in *this* case the distance between them is closing at *one and a half times* the speed of light. Which is three times faster than the speed at which the light is approaching the *front* door!

Remember we said both doors were the same distance from the light bulb. So if the distance between the beam of light and the *rear* door is closing three times faster than for the front door, that means (to the man on the embankment) the light is going to reach it sooner and cause the *rear* door to open before the *front* door!

Yet the railroad employee inside the coach saw both doors open at the same instant!

So we have a situation here where two men observe *the same event* to occur at *different times.* Yet neither one of them is wrong!

It's beginning to look like *time* is not as reliable as we thought.

The above is an example of how the *time* of an event (in this case the opening of the doors) may vary when observed from frames of reference moving at greatly different speeds.

It seems that time doesn't always pass at the same rate for everyone!

It is hoped that the matter has been sufficiently well explained, because if not then the rest of this book isn't going to make much sense.

Therefore if a reader is still not comfortable with the content of chapter seven, the chapter on the special relativity of light, then he should go to appendix (10). This appendix was added in order to further explain, this time with the aid of diagrams, exactly *why* one man sees the doors of the railroad coach open at the same time, while another man sees one of the doors open before the other.

CHAPTER EIGHT.

THE LORENTZ TRANSFORMATION.

What chapter seven has shown is that events *inside* the coach will *always* be seen the same way by a person *inside* the coach. *It makes no difference whether the coach is moving or standing still.* (Remember the man and his wife on the cruise ship on page 18.)

But to a man *outside* the coach if the coach is moving very fast *relative to him* then he will see things differently.

So if we *knew* what was happening *inside* the moving coach, is there a way to know how the man standing still would see it?

Earlier we mentioned a Dutch physicist by the name of Lorentz (Hendrik Antoon Lorentz, 1853-1928). He developed what's called the Lorentz Transformation. This is a series of equations by means of which we are able to *transform* an event inside the coach, and see how *the same event* would appear to a man on the embankment.

The Lorentz Transformation is the single greatest contribution to special relativity. (In this book we're simply going to be using what we'll call a "relativistic conversion factor" derived from the Lorentz equations.)

So far we've talked about the man on the embankment and the train. Instead let's think of them as "frames of reference."

You'll remember we first mentioned frames of reference back on page 17, and said you'd be hearing more about them.

There's a *moving* frame of reference (the train) and there's a *stationary* frame of reference (the man on the embankment). This will make it easier when talking about the Lorentz Transformation.

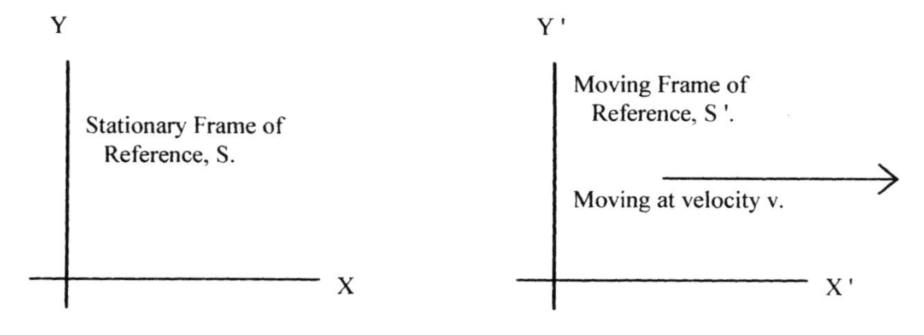

The stationary frame of reference (the embankment) we'll call S and the moving frame of reference (the train) we'll call S'.

The experiment with the train and the doors showed that time is not absolute, and that it can be affected by the relative velocities of the event, and of the person viewing it.

From the Lorentz Transformation we can derive the *relativistic conversion factor* that we just mentioned on the previous page. Then, knowing the details of an event taking place in the *moving* frame of reference, we can multiply (or sometimes divide) by this relativistic conversion factor to get details of the event *as seen by a person in the stationary frame of reference*. (Or, for that matter, the other way round!)

Now what follows is going to involve some math., especially in this chapter, and later in chapters thirteen and fourteen.

Since the invention of computers and hand held calculators nobody much uses even simple math. anymore in their daily lives. Some might therefore be intimidated by the explanations and the derivations of the various equations that form the basis of special relativity. But, as we said in the foreword, *the reader will still be able to gain a good understanding of the subject even without following all the math.* So don't be discouraged.

We will now try and come up with this relativistic conversion factor. For this we're going to need two clocks.

The clocks we're going to use are of a special type invented for this purpose by an American physicist called Feynman.* They are very good clocks. We can't say they keep perfect time because we've found out there's no such thing. But if we set each of these clocks to the same reading, and put them side by side on the kitchen table, they'll continue to agree for weeks and months and years. You might say they'll both "tell the same time."

Each clock consists of a light bulb, which gives out flashes of light, and a mirror which reflects these flashes of light back on to a counter which records their arrival and keeps count. The instant a flash is received the bulb flashes again, and the next flash of light sets out on its journey to the counter by way of the mirror.

On the next page is a sketch of one of these rather strange clocks.

* Not long before his death from cancer Richard Phillips Feynman (1918-1988) served on the panel investigating the spectacular 1986 failure of the space-shuttle Challenger. With particular emphasis on the part that may have been played in this disaster by the unusually low temperature prevailing at the time of the launch.

Mr. Feynman's remarkable clock.

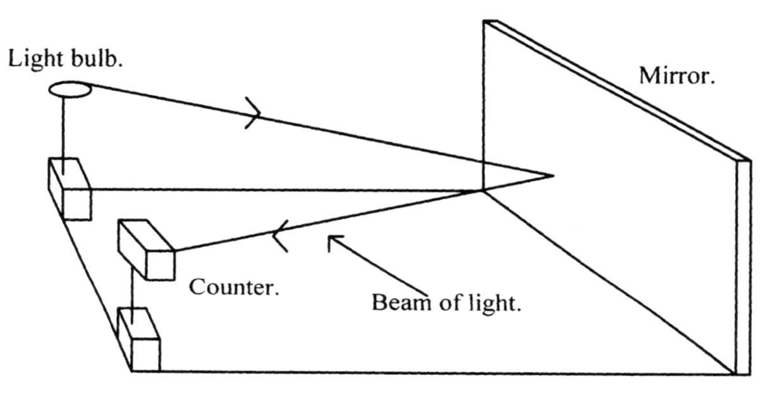

Every time a flash is received the counter will click once. The clock measures the passage of time by counting the number of clicks (just as old fashioned clocks used to measure time by counting the number of swings of a pendulum). As light *always* travels at the same speed, and provided it has the same distance to travel, the time between clicks will always be same.

Again, Feynman's clock measures time based, not on the swings of a pendulum, but on how long it takes, as seen by an observer, for the flash of light to travel from the light bulb to the counter!

We're going to put one of Mr. Feynman's clocks in each of the two *frames of reference*. Here is a picture of them in their frames of reference.

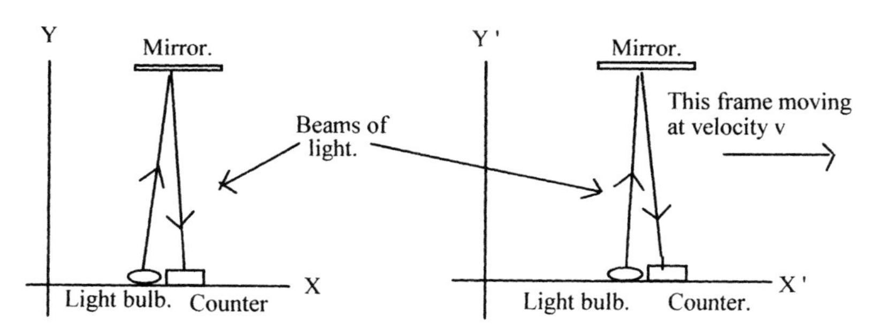

First (on the next page) we're going to consider the moving frame of reference (the train) and say there's a lady in it.

The moving frame of reference is moving very fast. The other frame of reference (the man on the embankment) let's say (for the sake of simplicity) that he's not moving at all. What the man *is* doing is looking at the *moving* frame.

The lady looks at her clock and sees it's behaving just as it was before they were moving. Had her clock been behaving differently then she'd have been able to tell she was moving without having to refer to any outside body. This would be a violation of Galileo's Principle of Relativity. (See chapter one.)

But the *stationary* observer, looking at the *lady's* clock in the *moving* frame of reference, he sees something different.

Let's consider the moving frame of reference, *as seen from the stationary frame of reference.*

Below is a diagram showing the *stationary* frame of reference S, and the *moving* frame of reference S' which (together with Mr. Feynman's clock) is moving very fast.

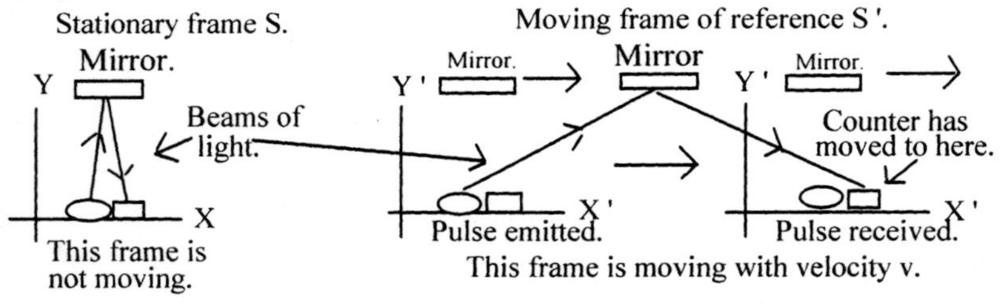

The man in the stationary frame of reference S is observing events in the moving frame S'. *As he sees it, in the moving frame the light has a lot farther to travel before it can catch up with the counter.*

Note that in the *moving* frame of reference, after the light has left the light bulb, *and while it's still on its way to the counter*, the whole frame *and everything in it* continues to move.

As seen from the stationary frame of reference, in the *moving* frame the light from the light bulb *has farther to travel*, because while the light is on its way to the counter, the counter meanwhile has been moving farther away so *the light has a greater distance to go* before it can catch up with it.

While the man and the lady both see the light move at the same speed (relative to *themselves,* of course) *as seen by the man in the stationary frame of reference* it will take longer for the light in the *moving frame* to reach the counter *because it has farther to go*! He'll hear fewer clicks from the *moving* clock than the lady does (because *to her* the light doesn't have so far to go). The man in the *stationary* frame of reference will see the *moving* clock *record time more slowly* than his own clock.

We're going to see how much is *the difference* in time between clicks of the clock in the *moving* frame, and between clicks of the similar clock in the *stationary* frame, both *as seen by* the man in the stationary frame. We'll use the following symbols:

a……will be the time between clicks in the *moving* frame.
b……will be the time between clicks in the *stationary* frame.
v……is the speed at which the *moving* frame is moving.
c……is the way we (always) denote the speed of light.*
d……is the distance moved by the counter while the light is trying to catch up with it.

(1) The time b between the clicks of *either* clock, as seen by a person in *the same frame of reference* as that clock, is determined by the speed of light, and by the distance traveled from the light bulb to the mirror, then from the mirror to the counter.

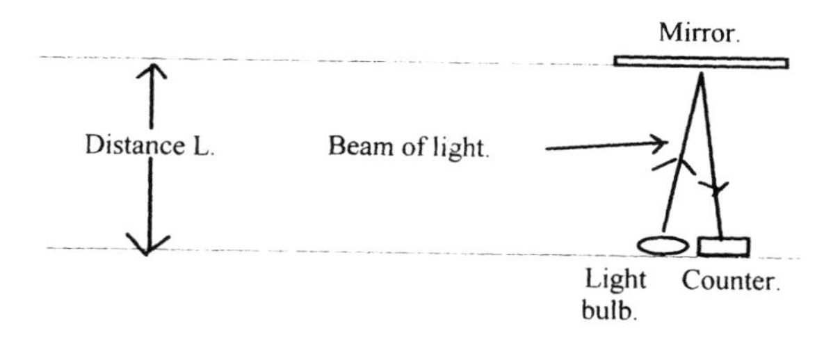

The distance the light travels is 2 times L and its speed is c.

Therefore the time taken is 2 x L divided by c. So $b = \frac{2L}{c}$ or $L = \frac{cb}{2}$.

* "c" was chosen as the international symbol for the speed of light because it stands for *celeritas*, the Latin word for swiftness.

(2) The time measured by the *moving* clock *as seen by the man in the stationary frame,* can be found using geometry. (Refer back to the sketch on page 58.)

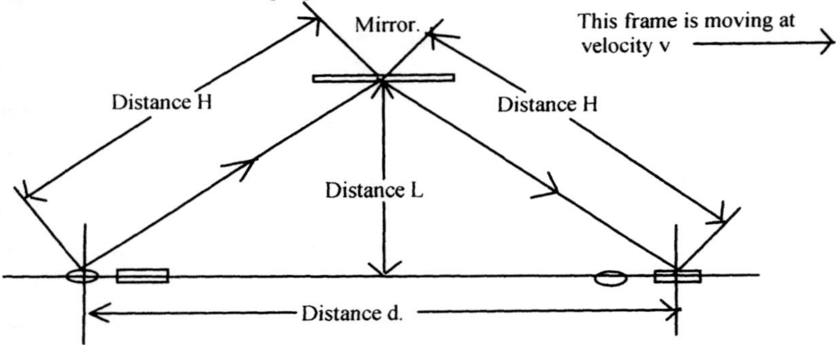

The time a, *seen by the man in the stationary frame of reference,* is the time taken by the light to travel one side of the triangle from the light bulb to the mirror, distance H, then another distance H from the mirror to the counter.

$$\text{So } a = \frac{2H}{c}. \quad \text{Or } H = \frac{ca}{2}.$$

Also, in the time a the moving frame travels a distance d, where d is the speed multiplied by the time. So d = v x a.

(3) Here's a slightly different version of the above diagram.

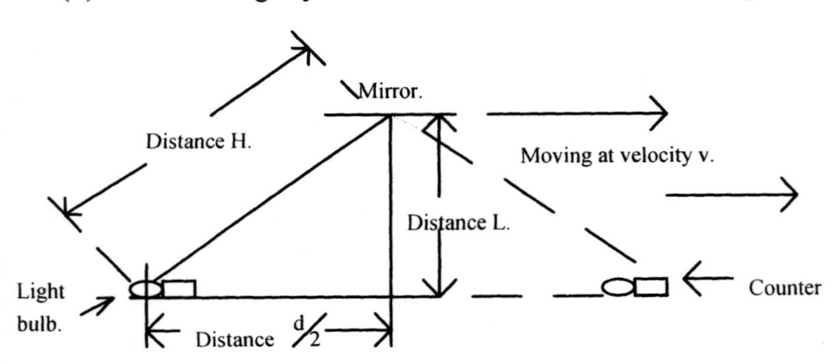

Now we've all heard of Mr. Pythagoras, of ancient Greece. He lived from about 580-500 B.C. (If you want to be politically correct you have to say B.C.E., which stands for "before the common era.")

(4) According to the aforementioned Mr. Pythagoras who, as we mentioned on the previous page, lived in ancient Greece,* the square on the hypotenuse of a right-angle triangle is equal to the sum of the squares on the other two sides.

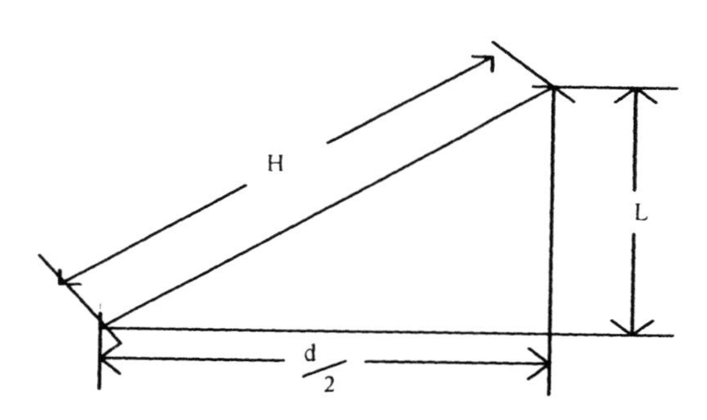

$$\text{Therefore } H^2 = (\frac{d}{2})^2 + L^2.$$

(5) From (1) above we know that $L = \frac{cb}{2}$

From (2) above we know that $H = \frac{ca}{2}$.

Also from (2) we know that $d = v \times a$, (or $\frac{d}{2} = \frac{va}{2}$.)

(6) In (4) above, we can substitute the values from (5) above so that for $H^2 = (\frac{d}{2})^2 + L^2$ we can write $(\frac{ca}{2})^2 = (\frac{va}{2})^2 + (\frac{cb}{2})^2$.

On the next page we're going to make use of this information that we've gathered.

* It wasn't ancient at the time, of course.

Now we'll solve to find a value for *a* in terms of *b*.

Starting with $(^{ca}/_{2})^2 = (^{va}/_{2})^2 + (^{cb}/_{2})^2$

and multiplying by 4 we get $(ca)^2 = (va)^2 + (cb)^2$

Dividing by c^2 we get $a^2 = (^{v^2}/_{c^2} \text{ x } a^2) + b^2$

which means $1 = ^{v^2}/_{c^2} + ^{b^2}/_{a^2}$

or $1 - ^{v^2}/_{c^2} = ^{b^2}/_{a^2}$

therefore $a^2(1 - ^{v^2}/_{c^2}) = b^2$

and transposing we get $$a^2 = \frac{b^2}{(1 - ^{v^2}/_{c^2})}$$

And taking the square root we get $$a = \frac{b}{\sqrt{1 - ^{v^2}/_{c^2}}}.$$

This means the man in the stationary frame of reference sees *a longer period of time* elapse between clicks of the clock in the moving frame of reference than between clicks of his own clock!

The clock in the moving frame of reference is recording time more slowly! That means to the man in the stationary frame of reference, while his own clock moves forward an hour, say, then *during that same period*, as he sees it, time will have advanced *less than an hour* by the clock in the moving frame of reference.

Let's switch this around and think instead only of *how fast time is passing* in each of the frames of reference. For if there's a longer period of time between the clicks of the clock in the *moving* frame of reference that means time there must be passing *more slowly.* While a given event takes place *more time will be passing in the stationary frame of reference than the moving frame.*

Let's use the symbol t to denote the amount of time passing as recorded by the clock in the *stationary* frame of reference, and t' to denote the amount of time passing as recorded by a similar clock in the *moving* frame, *both as seen from the stationary frame!*

From what we just said above we know that t must always be equal to, or greater than, t' (so we know which one is bigger).

And we've already established the relationship between the rates at which time is seen to be passing in the two separate frames of reference, when observed by a person in *one* of those frames of reference (it doesn't matter which one). So knowing these things, and re-arranging the above equation, we have the following:

Taking $a = \dfrac{b}{\sqrt{1 - v^2/c^2}}$ we can modify it to $t = \dfrac{t'}{\sqrt{1 - v^2/c^2}}$.

This establishes the relationship between *the rate at which time is seen to be passing* in a moving frame of reference (which we said we'd call t') on the one hand, and *the rate at which time is seen to be passing* as measured by a clock in the observer's own stationary frame of reference (which we said we'd call t) on the other. *Both as seen from a stationary frame of reference*

Depending on the velocity of the moving frame of reference the rates at which time is seen to be passing will be different!

Note the expression $\sqrt{1 - v^2/c^2}$. Another way to write it would be $(1 - v^2/c^2)^{1/2}$. You'll be seing this a lot because it's the relativistic conversion factor that we first mentioned back on page 55. It's a value obtained from the Lorentz Transformation and is the key to *transforming* events in a *moving* frame of reference, to show how *the same event* would be seen from a *stationary* frame of reference.

Normally the value of the relativistic conversion factor is about equal to one, but as velocity v nears the speed of light the factor becomes very small, ultimately approaching zero. This results in the *same event* being seen *differently* by a person in the *stationary* frame of reference and a person in the *moving* frame of reference.

It is the Lorentz Transformation that defines the relationship between the two frames of reference.

What we've found out in this chapter is that the rate at which time passes, *previously thought to be fixed and unalterable,* is in fact variable and *depends on the relative velocities of the event, and of the person viewing it.*

In the next chapter we'll go on to show how this distortion of time can even affect the rate at which a person grows old!*

* When two people get married it's usually with the expectation of growing old together. But according to special relativity this doesn't necessarily always have to be the case! As we shall see in chapter nine.

CHAPTER NINE

HOW THE TIME OF AN EVENT CAN VARY.

"I don't know what you mean," said Alice.

"Of course you don't," said the Mad Hatter, tossing his head contemptuously. "I dare say you never even spoke to Time."

"Perhaps not," Alice cautiously replied, "but I know I have to beat time when I learn music."

"Ah, that accounts for it," said the Mad Hatter. "He won't stand beating. Now if you only kept on good terms with him he'd do almost anything you liked with the clock. For instance, suppose it were nine o'clock in the morning, just time to begin lessons. You'd only have to whisper a hint to Time and round goes the clock in a twinkling! Half-past one, time for dinner!"

"I wish it was," said the March Hare to itself in a whisper.

Lewis Carroll.
Alice's Adventures in Wonderland. 1865.

The Mad Hatter knows that clocks don't always keep the same time? Let's see what this means.

Say an astronaut travels out into space, then turns around and comes back. He travels at eight-tenths the speed of light and the trip takes twenty-five years. He leaves behind a twin brother. At the start of the trip the twin brothers were twenty years old.

But when the astronaut returns he's surprised to find he's now much younger than the twin brother who'd stayed behind on earth.

How much younger *was* the astronaut? We'll use the equation from page 63.

$$t = \frac{t'}{\sqrt{1 - \frac{v^2}{c^2}}}$$

To the twin back on earth, by *his* clock the trip took twenty-five years. So in the equation the value of t will be 25 years.

But during the trip *time* in the capsule, *as seen from earth,* was passing more slowly! The moving clock *as seen from back here on earth* was clicking more slowly. (Refer back to the description of Feynman's clock, starting on page 56.)

In the equation on the next page this *shorter* time will be denoted by t'. We also know that v was 80% of the speed of light.

Transposing the above equation we get $t' = t\sqrt{1 - v^2/c^2}$.

Substituting, $t' = 25\sqrt{1 - (0{\cdot}8c)^2/c^2} = 25\sqrt{1 - (0{\cdot}8)^2}$

$= 25\sqrt{1 - 0{\cdot}64} = 25\sqrt{0{\cdot}36} = 25 \times 0{\cdot}6$ $\underline{= 15 \text{ years}}$

Now we said that when the astronaut brother left to go on his trip the twin brothers were (each) twenty years old. Which means that when the astronaut returned he was thirty-five while his twin brother who'd remained behind on earth was now *ten years older!*

The astronaut's clock *as seen by him* doesn't slow down. But because the earth (and his turn-around point, and everything in *that* frame of reference) are moving at high speed *relative to him on his outward journey, and again on the return journey*, then the astronaut sees *distances* in that frame of reference *shrink* (which is explained in chapter ten). Consequently it takes *less time* for the astronaut to cover these *shorter distances*. It takes him in fact only *fifteen* years to make the whole trip, while his twin brother who remained behind on earth had to wait *twenty-five* years for the astronaut to return.

Now you may think you've found a flaw in this argument, and why was it the *twin* who aged faster? Why couldn't it just as easily have been the other way around? According to Galileo it would have been equally correct to consider the *twin* as traveling relative to the *astronaut!* Good point, see appendix (4) for more on this.

The above shows how time is not absolute and that *time* in a *moving* frame of reference will be seen to be passing *more slowly* by someone in a *stationary* frame of reference. *It's the speed of light that's absolute now, and everything else has to fit in with this.*

But it's not only time that changes when things are traveling at high speed. We're going to show that *distance* isn't any better. You can't rely on *distance* (or length) any more than you can on *time!*

After all, what is distance? When you're driving in your car, distance is the *speed* at which you're traveling multiplied by the *time* you continue to drive at that speed. And we've just found out that one of these, *time,* can't be relied upon!

In chapter ten we'll show how *the length* of a body moving at high speed (a body in a *moving* frame of reference) will appear to be *shorter* viewed by someone who's *not* moving (in a *stationary* frame of reference) than it would to someone moving along with the object.

CHAPTER TEN.

YOU CAN'T RELY ON DISTANCE EITHER.

Let's go back to that railroad coach which was traveling at high speed. Imagine the man in the middle of the coach gets up and walks to the front door. How far has he gone?

To anyone inside the coach he's gone half the length of the coach. But to the man sitting on the embankment watching the train rush past the man inside the coach has gone farther. (After all, the train is moving too.)

Consider how far the man inside the coach has walked *relative to the coach* as seen by (a) a person *inside* the coach and (b) the man watching from the embankment.

The *speed* at which the man is walking relative to the coach *is the same no matter who's watching*. (Special relativity doesn't have anything to do with the speed at which a person is observed to be walking!) But *the rate at which time passes* is not the same. (We just came to that conclusion in chapter nine, didn't we?)

So if the *speed* at which the man inside the coach walks is the same to all observers, and a shorter *time* has elapsed inside the coach as seen by the man on the embankment, then *to him* the distance walked by the man in the coach must be shorter.

We're able to calculate the relationship between the distances by using the Lorentz Transformation equations.

Now if we wanted to measure the length of the coach how would we do it? For the man *inside* the coach it would be easy. All he'd have to do would be to take a tape measure and measure it. But for the man on the embankment, to determine the length of the coach, *as seen by him*, he'd have to mark the positions on the embankment that were being passed by the front of the coach, and by the rear of the coach, at the *same time as observed from the embankment*, then measure the distance between these two marks.

According to special relativity, the length of the railroad coach, *as measured from the embankment*, would be different from the length that would be obtained *by measuring the train itself.*

In point of fact, in line with what we just said on the previous page, because of the unreliability of time, the length of the coach as seen by someone sitting on the embankment, would be *shorter* than it would be to a person traveling along with the coach!

To the man on the embankment, the time that elapses while the coach is passing him, *as measured by a clock he's able to observe inside the coach* (if this were possible) will be *shorter* than the time taken for the coach to pass *as measured by his own clock.*

Because, to the man on the embankment, less *time* has passed in the moving frame of reference, and because the *length* (distance) of the coach is its *speed* multiplied by the *time* it takes to pass, then *to him* (to the man on the embankment) the coach must appear to be *shorter.*

Instead of the train this time let's consider the case of a simple iron bar, say one foot long. This bar is traveling, in the direction of its length, at *half* the speed of light.

Special relativity says the *contracted* length of the iron bar is proportional to $\sqrt{1-v^2/c^2}$, again where v is the speed of the iron bar and c is the speed of light.

If the speed doesn't change, but the *time* is reduced, *then so must the length be reduced by a proportional amount.*

Let L be the length of the iron bar when it wasn't moving (which is also the length that would be seen by a person moving along with it). And let L' be the length of the bar when moving *as seen by the person who is not moving.*

We'll use the relativistic conversion factor again.

$$L' = L\sqrt{1-v^2/c^2} = L\sqrt{1-(½)^2}$$

$$= L\sqrt{1-¼} = L\sqrt{0\cdot 75}$$

$$\underline{L' = 0\cdot 866\ L}$$

This means to anyone watching the iron bar as it whizzed by it would appear to be *shorter*, and the foot long bar would seem to be only 0·866 of a foot, or a little over $10^3/_8$ inches long. (You may remember we used this as an example on page 7.)

Of course for this the iron bar has to be traveling at half the speed of light, which would mean traveling at about 93,000 miles per second.

By everyday standards down here on earth this is ridiculous. To launch a satellite into orbit requires a speed of only about five miles per second. To escape from the earth's gravity altogether, as in going to the moon, we need only achieve a speed less than seven miles a second, and to do this we need a rocket the size of a New York skyscraper almost! To achieve a speed of 93,000 miles per second we'd need a rocket *millions* of times bigger. So big that, when it blasted off, it would incinerate everything for hundreds of miles around!*

The above demonstrates that objects traveling at speeds near to the speed of light, *relative to a stationary observer,* will appear *to that observer* to be *shorter* in the direction of travel.

For an example, let's take the case of a police car traveling at a very high speed (a speed approaching the speed of light).

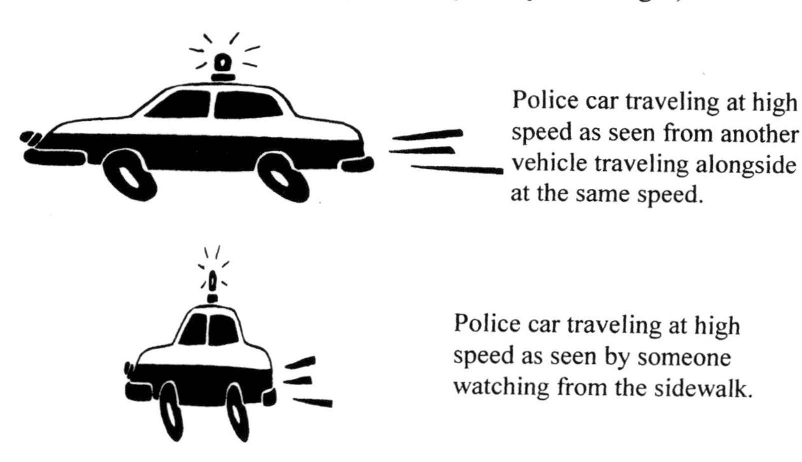

Police car traveling at high speed as seen from another vehicle traveling alongside at the same speed.

Police car traveling at high speed as seen by someone watching from the sidewalk.

The police car, traveling at extremely high speed, will be seen to shrink, in the direction of travel, when observed from a stationary frame of reference.

* See appendix (8).

So, to sum up!

We told how, by means of experiments such as those carried out by the Americans Michelson and Morley, and earlier by people like Maxwell and Mach (see footnote on page 45) it was found that the speed of light *is always the same*, that it is *absolute*, and that everything else has to fit in with this.

Then we showed how, as a result of this, *time* may be seen to pass at a different rate by one person than by another.

Now we've shown how not only time, but *distance* (or length) may also be seen differently according to the relative velocities of the event, and of the person observing it.

But that's not the end of it. There's still one more thing that changes near the speed of light, and that is *mass!* At least in one way it changes, yet in another way it doesn't! (We'll try and explain what we mean in the next chapter.)

We talked a bit about *mass* in chapter five. Now we're going to take a closer look at it in chapter eleven.

CHAPTER ELEVEN.

MASS CAN ALSO CHANGE.

If the Lorentz transformation is to be believed then nothing can travel faster than the speed of light. This is because, according to Lorentz *time*, in a frame of reference moving at the speed of light (as seen by a stationary observer) would *stop altogether*, and the *length* of an object in that moving frame reduce to *zero*. And that raises the question of what would happen if the body were moving *faster* than the speed of light. Could the length of anything be *less* than zero?

Einstein concluded the answer is that *nothing in the universe can ever travel faster than the speed of light*.

But what was to stop us from accelerating a body past the speed of light if we wanted to?

In accelerating a body what we have to overcome is the body's *inertia,* its resistance to being moved. We come across this all the time. It's harder to push a city bus than a shopping cart. Its *inertia* is greater. And near the speed of light, Einstein said, this inertia must become huge, eventually approaching infinity. It will become so great that ultimately *the body just won't go any faster no matter how hard you push it!* The result is the speed of a body might *approach*, but can never quite *attain*, the speed of light.

But a body's inertia depends on its mass! Everything points to the two being directly related. Newton obviously thought they were! So when a body is being accelerated (relative to an observer) then if its *inertia* is increasing that means its *mass* must be increasing!

Yet we've always thought of a body's mass as fundamental to its very being. How many atoms it has, how many neutrons and protons, that sort of thing. So how can it increase?

Well it's true that physicists will tell you a body's mass *never* changes! (They produce all kind of mathematics to prove it.) This fixed mass is the body's real mass. At one time it was referred to as the *proper* mass, and the other kind, the kind of mass that changes, as the *non*-proper mass.

But in this book we're going to use another name. The mass that increases with a body's relative speed we're going to call the *inertial* mass. (It's the name that Einstein used, so we're in good company.)

So from now on, when we talk about a body's mass *increasing*, remember what we mean is its *inertial* mass. (That way we'll keep the physicists happy.)

On the previous page we said that because a body's speed can never exceed the speed of light, the body's *inertia* (its *inertial mass*) must become huge, eventually infinite!

But if we keep pushing the body, and because of the huge inertia it's hardly going any faster at all, then *where is all the energy going?*

Newton said that a moving body possessed *energy* simply by virtue of the fact that it was moving. But apart from the knowledge that a moving body could cause a lot of damage if it hit something, what *other* evidence of this energy did we have? Where was it now, and shouldn't there be some way to measure it?

So all this energy must have gone *somewhere* but we just don't know *where*.

On the other hand we said the (inertial) mass of the body has increased, which raises the question of where has this extra mass come from! The old law of conservation of mass (chapter five) says it can't just appear out of nowhere!

Now we have some *energy* that has disappeared, and can find *no evidence* of, and some (inertial) *mass* that has apparently *come from out of nowhere.*

It was then that Albert had a flash of inspiration!

The energy applied to make the body go faster *must instead* have gone to increase the body's inertial mass! *And if energy can turn into mass they must be two forms of the same thing.*

If *mass* and *energy* are two forms of the same thing, and can be changed from the one to the other, then the old laws of conservation of mass and conservation of energy can't be right!

Remember we told you (pages 37-38) how everyone had always thought that *individually* the total energy, and the total mass, were each of them constant and could never increase or decrease.

Now it seems that, despite what we've always thought, the total amount of energy can't be a constant. Neither can the total amount of mass be a constant. For if one of them can be changed into the other (examples of which we're going to show you further on) then neither one of them can *individually* be considered constant.

The only reason people continued to believe for such a long time that the overall quantity of mass never changed was because the increase in *mass* resulting from the application of *energy* was too small a portion of the total mass to be measurable back then.

Now we must find a new law to replace the old ones. The new law says *the total of mass plus energy must always be the same!*

Individually mass and energy may vary, but the *sum total of the two together* can't *ever* get any bigger or any smaller.

This was going to be the new law! Whatever energy and mass exists today was evidently contained in the *singularity* that suddenly and mysteriously appeared some roughly fourteen thousand-million years ago. It's thought this quantity is fixed and not likely to change. (See also pages 116 & 117 where we talk about dark energy.)

It was this understanding of special relativity that led to the realization that *energy* can change into *mass*, that they must be *just different forms of the same thing*. Which means the reverse ought to be true and that it should be possible to turn *mass* into *energy*. This led to the development of the atom bomb, and later to its big brother the fusion bomb, or thermonuclear device, which will probably bring about the extinction of the human race sooner or later.

One thermonuclear test carried out by the former Soviet Union in nineteen sixty-one involved the release of energy equivalent to some *fifty-seven million tons* of trinitrotoluene (TNT). The resulting fireball was said to have been thirty miles in diameter.*

For much of the cold war four thermonuclear bombs, thought to be of twenty megatons each, were routinely carried aboard B-52 bombers of the U.S. Air Force. There were some five hundred of these bombers available, sixty-six of them being in the air at any given time and ready to launch an immediate attack upon the Soviet Union.

* This was in fact a smaller version of the *one hundred* megaton bomb that had originally been planned. That monster would have been an effective weapon of a size and weight that could have been delivered by aircraft available to the Soviet Union at the time.

So now we've arrived at the conclusion that *mass* and *energy* are just two different forms of the same thing, and can be changed from the one into the other. This relationship between mass and energy is given by Einstein's now famous equation $E=mc^2$ which we're going to talk more about later.

Meanwhile we're going to consider some of the applications of this interchangeability between mass and energy. We're going to take a look at how nuclear energy is being used here on earth, in many cases to our benefit. And we're going to take a special look at how nuclear energy causes the sun to be hot.

Because if the sun wasn't hot, and if some of the sun's mass wasn't continuously being changed into energy all the time, then it's a sure thing none of us would even be here!

Nuclear energy, it seems, is not only responsible for our very existence, but will sooner or later be the agent of our demise! If we don't arrange our own cremation first the sun is eventually going to do it for us. (See pages 79 and 80.)

Changing mass into energy is what we're going to talk about next, in chapter twelve.

MORE ABOUT ENERGY AND MASS.

The great breakthrough in man's attempt to harness the energy of the atom came when the nucleus of an atom of uranium was split.

The rare isotope U-235 is the only *naturally occurring* fissile material in which a chain reaction can be sustained. It accounts for only about seven-tenths of one percent of all uranium. Nearly all the rest is U-238, which is normally not fissile. But scientists working independently at the University of California were able to produce *another* fissionable element, plutonium 239 (see page 171).

Scientists around that time had been bombarding atomic nuclei with neutrons with varying results. Neutrons, having no electrical charge, would not be repelled by the positively charged protons in the target material's nucleus. Since neutrons were only identified in 1932 alpha particles (actually helium nuclei, now known to consist of two protons and two neutrons) had been used prior to this.

In 1919 Ernest Rutherford (1871-1937) by bombarding atoms of nitrogen with alpha particles succeeded in changing them into atoms of oxygen. (This was not *fission*, however.)

Then in 1938 two German scientist, Otto Hahn (1879-1968) and Fritz Strassmann, while bombarding nuclei of U-235 with neutrons found that traces of the element barium had appeared. Barium is only about *half* the atomic mass of uranium. It seemed that a neutron must have hit a U-235 nucleus and *split it in two* (p.179-180).

As this fission also yielded a net gain of two or three unattached neutrons, something that had been forecast by Leo Szilard (see page 180) *the possibility existed* that they could go on to cause fission in *other* U-235 nuclei, producing *more* free neutrons, and so on.

After the atom had split the mass of the pieces left over was *less* than they'd started with! *The missing mass had turned into energy.*

We said that several neutrons were *also* produced, which *could* go on to split other U-235 nuclei. (This is not the case with U-238, for example.) These neutrons, before they can escape to the outside, *may or may not* hit other U-235 nuclei. But if the piece of U-235 is big enough, and constitutes what's known as a critical mass (see next page) then the chance of the stray neutrons hitting *other* U-235 nuclei reaches the required probability threshold. The resulting fissions release *still further* neutrons, which go on to hit still *other* nuclei, and so on and so on, causing a chain reaction. Then, *in an instant,* all of the U-235 nuclei in the critical mass will fission.

For the *least amount* of fissile material to make a critical mass it would have to be in the shape of a sphere. (For U-235 this is thought to be about fifteen pounds.) *Any other shape* would have a greater surface area through which neutrons could escape before they had a chance to hit other U-235 nuclei (so you'd need to have more of it).

It's not difficult to make a uranium bomb. All you need to do is bring two pieces of *extremely pure* U-235 together *fast enough*, to make a critical mass, at the same time introducing a source of stray neutrons at just the right moment. Then it will go off all by itself!

Einstein didn't believe that energy could ever be obtained from the atom in practice. It was an interesting theoretical concept, he said, nothing more. Splitting an atomic nucleus, Einstein declared, was "*like shooting birds in the dark, in a country in which there are very few birds*." The noted British scientist Frederick Lindemann agreed. God, he said, knowing how irresponsible we were, would never permit such awesome power to fall into our hands.

The difficulty in making a uranium bomb is in separating out the U-235 from the U-238. There are four principal ways to do this. One way is by electromagnetic separation, while another method used is thermal diffusion. Centrifugal separation is a method we hear a lot about these days, from places like Iran and North Korea. And finally there is gaseous diffusion, where the metal is converted into a highly corrosive gas uranium-hexafluoride. This is thought to have been the method used to extract the U-235 which underwent fission above the Japanese City of Hiroshima on the morning of August 6th, 1945.

Separation is achieved by making use of the slightly differing masses of the uranium isotopes. However, all known methods of enriching natural uranium (increasing the proportion of U-235) are costly, complicated and time consuming. Which is probably a good thing, otherwise everyone would have atom bombs.

The device used against the city of Hiroshima (see above) took all the U-235 we had at the time. We couldn't have tested it first, since then we'd have had no U-235 left to make another one with!

Plutonium is easier to obtain than U-235 and its fission releases more energy. But initiating a chain reaction in it is more difficult.

Because nobody was sure if it would even work the plutonium bomb was tested in the New Mexico desert early on July 16th, 1945.

It was another plutonium bomb that destroyed the Japanese city of Nagasaki on Aug. 9th, and there was still enough plutonium left for two or three more similar bombs should they have been needed.

The United States now having a monopoly of such weapons, they were then brandished before the Soviet Union, which had just joined the Pacific war and had the world's mightiest army, in the furtherance of American foreign policy. Specifically in proposing to the said Soviet Union that she abandon her plans to invade Japan, a move the U.S. would otherwise have been powerless to prevent.

The U-235 bomb ("little boy") that demolished the Japanese city of Hiroshima was the equivalent of about 13,500 tons of the high explosive TNT. Hearing about it on the radio Einstein said if he'd known they were going to do this he'd have become a shoemaker.

The plutonium bomb ("fat boy") was more powerful and had the explosive energy of about 22,000 tons of TNT. At the time both the uranium and the plutonium bombs were known as atom bombs.

Today we use the term nuclear rather than atomic, because it's from the *nucleus* of the atom that we obtain energy.

The problem with nuclear fission is that it produces radiation, though this is nothing compared to the massive pollution caused by burning vast amounts of precious and irreplaceable fossil fuels.

This radiation consists mostly of alpha particles, beta particles, and gamma rays. Alpha particles (which we said are helium nuclei) can't penetrate a piece of paper even. Beta particles (which are free electrons) are more deadly, but even they can't penetrate your skin. They are harmful if they're in something you eat or drink, or especially if you inhale contaminated particles.

It's gamma rays (see page 26) that are the problem with nuclear fission. They are *waves*, a transmission of energy, and didn't used to be thought of as particles, though the distinction between particles and waves is no longer as clear as it once was.

As an alternative to the splitting of *heavy* atoms, energy can also be obtained by fusing *lighter* atoms into *heavier* ones.

The simplest and still the most abundant element in the universe is hydrogen. It was the first element to be formed, or at least one of the first two. The other one, the next simplest element, is an inert gas called helium, a name derived from *helios*, the Greek word for sun. This is because helium was first discovered *in the sun* by spectrographic analysis of light from the sun, where certain dark lines in the spectrum revealed an hitherto unknown element (see page 130).

On earth helium is found in natural gas deposits, particularly in the states of Texas and Kansas, where it results from the radioactive decay of isotopes of elements like uranium and radon.

The USA is the only place where helium has ever been available in commercial quantities. Her reluctance to export the gas, and the consequent need by other countries to use the highly combustible hydrogen in its stead, was a major factor in hastening the demise of the dirigible airship as a viable form of passenger transport. (One recalls the explosion of the airship Hindenburg in May of 1937.)

For centuries it had been a mystery where the sun's seemingly inexhaustible supply of energy came from. In 1938, when this author was in high school, Hans Bethe and Carl Friedrich von Weizsacker said they thought they knew what was the source of all the energy. It came from the fusion of hydrogen into helium, they said.

We now know that in the sun some *six-hundred-million tons* of hydrogen are being consumed *every second!* (Other sources put it at somewhat less.) As hydrogen weighs only one fifteenth as much as air this is an incredible amount. This is *every second,* remember, and it's been going on now for almost five *thousand-million* years!

If *all* the hydrogen was converted to energy, instead of most of it fusing into helium, the sun would only need to be burning between four and five million tons of it per second instead of the (up to) six *hundred* million tons being consumed now.

In dealing with stars (the sun is a star) we deal with numbers impossible to imagine. A million earths would fit inside our sun, yet as stars go it's not all that big! The Latin word for star is *astra,* so when we talk about such big numbers we say they're *astra*-nomical. Strictly speaking astronomy is the study of *stars!* That is what it *literally* means (to use the word correctly for once).

It takes *four* hydrogen atoms to produce *one* helium atom. On a recent television program the narrator (a well known actress) said it takes *two* atoms of hydrogen to fuse into *one* atom of helium. This is not possible, however.

The atomic mass of an atom of hydrogen is 1·007825035, so two of them would weigh 2·015650070. But the atomic mass of an atom of helium is 4·00260324. (In both cases these figures are for the most common isotopes.) So clearly *two* atoms of hydrogen are not going to be enough to make a helium atom.

A hydrogen atom's nucleus consists of a single proton. What happens is the *four* protons from the *four* nuclei join up to become a *single* helium nucleus consisting of two protons and two neutrons. And if there are now only two protons then only two electrons are needed. (All stable atoms contain an equal number of each.)

So two of the (positively charged) protons have joined with two of the (negatively charged) electrons, canceling each other out and turning into neutrons, which have no charge! But the *mass* of what we have now *is less than what we had when we started!*

We said the atomic mass of a hydrogen atom is 1·007825035, so that *four* of them would weigh 4·031300140. If the atomic mass of an atom of helium is 4·00260324 that means an amount of 0·0286969 has *disappeared.* The missing bit of mass is called the *mass defect.* We now know it's been changed into *energy* (each new helium nucleus producing one quantum of energy).

If the sun is burning *six-hundred-million tons* of hydrogen per *second,* or *fifty-two trillion,* that is to say fifty-two million-million or (52×10^{12}) tons per day, you'd think by now, after nearly five *billion* years, it would be starting to, quite literally, run out of gas. (At least as far as *hydrogen* gas is concerned.)

Well even after all this time the sun is still made up of about 74% hydrogen and 23% helium* (here, too, estimates vary slightly) and at the current rate wouldn't actually run out of hydrogen for another ten or twenty *billion* more years. But long before this happens, in only about *five* billion years from now, we're going to have a problem.

The sun at that time will have become so hot that the oceans will boil away and all life will come to an end here on earth. And for that matter anywhere else in the solar system where life might exist, with the possible exception of (what used to be a planet) Pluto, which is far enough away from the sun that, who knows, conditions there might turn out to be just right. Like Waikiki Beach maybe!

* The rest of the sun is made up of heavier elements such as carbon, nitrogen, oxygen, aluminum, magnesium, silicon and iron etc., in fact all the known stable elements. At the temperatures inside the sun all exist in the form of gases.

The trouble is that the fusion of hydrogen into helium requires a temperature of about twenty-eight *million* degrees Fahrenheit, or sixteen million Celsius, and it's only at the *center* of the sun where temperatures are *that* high! And in the center of the sun already half the hydrogen has been used up, so that the sun is now about halfway through its life. Incidentally the sun's core accounts for about fifty percent of the total mass of the sun, but only about two percent of its volume.

In only about five thousand-million years from now virtually all of the hydrogen in the sun's core will have fused into helium. When this happens the sun, just temporarily, for only a few hundred million years or so, will swell to from one hundred to two hundred times its present size, fueled now by the fusion of helium into carbon (three helium nuclei fusing into one of carbon). Though now *cooler*, its enormous surface area will make the giant sun a thousand times brighter than it is today. Due to its relatively small *mass*, however, our sun will not progress beyond this helium burning stage.*

By then the matriarch of the solar system will have become a *red giant*, swallowing up Mercury and Venus and burning the earth to a cinder. The only reason why the earth itself probably won't end up *inside* the sun is because the sun's mass by then will have shrunk (by about 25%) so that all the bodies that orbit around her, together with *their* satellites, will have drifted farther away.

We said it's only at the center of the sun where it's hot enough for fusion of hydrogen to occur. It would be hard to reproduce these conditions here on earth, and even if we could do so there probably wouldn't be anyone left alive afterwards to record the results.

Fortunately (if that's the right word) there are two *isotopes* of hydrogen that we can use to make (so called) hydrogen bombs since they fuse into helium at lower temperatures than ordinary hydrogen. (Isotopes are simply variants of the same element and are chemically indistinguishable from one another.) In the case of hydrogen these are *deuterium* which, as the name suggests, has an atomic mass about twice that of the more common isotope (2·014101779 as compared to 1·007825035) while the even rarer *tritium* (you guessed it) has an atomic mass approximately three times greater (3·01605).

* In the larger stars heavy elements get fused into *still heavier* ones, with increasing numbers of protons, up to the element iron (# 26). Elements heavier than iron are produced in supernova explosions.

The simplest hydrogen isotope H-1 (or ^{1}H) has just one *proton* in its nucleus and a single *electron* whizzing round it.

As for the heavier hydrogen isotopes, a deuterium atom H-2 (or ^{2}H)* has one *proton* and one *neutron* in its nucleus while a tritium atom's nucleus, H-3 (^{3}H) has one proton and *two* neutrons.

Deuterium is sometimes called heavy hydrogen. Water, but with deuterium in place of hydrogen, is commonly known as heavy water.

The mass of an electron is so small it's usually ignored. (A neutron, for example, has 1,840 times the mass of an electron.)

H-1, H-2 & H-3 are also the names of Hawaii's three freeways (part of the so-called "interstate" system) on the island of Oahu (it means *gathering place*) though they probably weren't named for the isotopes of hydrogen.

It's the heavier isotopes of hydrogen, deuterium and (indirectly) tritium, that are used here on earth, unlike what's taking place inside the sun, to produce nuclear fusion, the plan evidently being to do this in the proximity of people we don't get along with.

Tritium is very rare. It is a radioactive isotope with a half-life of 12·26 years. It is produced by cosmic rays which, coming from outer space, bombard the upper reaches of the atmosphere causing nuclear reactions that create tritium.

Protons and neutrons are of *about* the same mass (not *exactly*) and it is their *combined* number (the total number of protons *and* neutrons) that determines an atom's designation (e.g U-*235* or *238*).

While uranium, with 92 protons in its nucleus, is the heaviest naturally occurring element (in any significant quantity) a number of even heavier elements can be produced artificially.

There are nineteen of these so far, possibly twenty-five, which you'll find listed in appendix (12). The best known is the fissionable element plutonium, virtually all of which has to be made by man.

* Actually deuterium has its own symbol, D, though it's really just an isotope of hydrogen.

Protons have a positive charge. And on page 78 we said atoms normally contain negative charged *electrons* in equal number to the number of protons. Neutrons, as the name suggests, have no charge.

Remember we said it's only in the *center* of the sun where the temperature is high enough for hydrogen atoms to fuse into helium. So far the only way we've been able to start a fusion reaction here on earth is by using a fission bomb as a detonator (see below). To do this in a laboratory wouldn't leave much of the laboratory (or of the surrounding city) so until we can find a better way of doing it the use of nuclear *fusion* for peaceful purposes seems as far away as ever!

Hydrogen bombs are mostly *fission* bombs anyway. A uranium (or plutonium) bomb is placed in a casing of lithium deuteride (this being an alloy containing deuterium). The resulting fission produces neutrons which bombard the lithium and produce *tritium* ions, which *then* mix with the deuterium that we just mentioned. The energy liberated in the fission process raises the temperature and pressure of the hydrogen isotopes to extremely high levels and converts them to plasma particles, which then undergo *fusion* into helium atoms with the almost instantaneous liberation of enormous amounts of energy.

Meanwhile the whole of the apparatus has been encased in a big jacket of the cheaper and far more readily available U-238. *Now*, with fusion of the hydrogen isotopes, the temperature is so high that even *this* most common form of uranium will undergo fission.

The fusion reaction is clean, but the two fission reactions (which account for most of the energy) produce high levels of radioactivity.

The atomic bombs delivered to the Japanese cities of Hiroshima and Nagasaki in Aug.1945 are the only occasions on which nuclear weapons have been used against human beings. "We didn't need to use that awful thing," said General Eisenhower, a view shared by eight of the nine American officers to have ever worn five-stars (the ninth being a general then serving in the president's cabinet, his view on the bomb's use no doubt influenced by political considerations).

Still it does seem that use of the atomic bomb accounts for why Japan today does not consist of a communist north (a rogue state like North Korea perhaps) and a democratic south. America's possession of the bomb, *and her demonstrated readiness to use it*, caused the Soviets to call off their planned invasion of the Japanese mainland. Consequently they were able to occupy only the four small Kuril Islands, what Japan calls her northern territories, and which Russia still holds to this day (see top of page 77).

CHAPTER THIRTEEN.

BRINGING NEWTON'S LAWS UP TO DATE.

Most of modern science and engineering is based on principles set out and published in 1687 by Sir Isaac Newton.

Newton

These principles went unchallenged for more than two hundred years, and are still in use world wide today. Engineering students apply Newton's laws constantly. They are the basis for just about everything he* studies. Yet special relativity, and the resulting need to modify those laws, is seldom encountered.

However, special relativity has shown us that when dealing with velocities near the speed of light Newton's laws are no longer valid.

As we seldom have to deal with things going anywhere near the speed of light this doesn't matter much to most of us. Even so, the purpose of this book is to show how special relativity alters things, so now we're going to see how Newton's laws had to be modified (not thrown out altogether) to bring them into the twentieth century.

We talked about what *mass* really is, and how in order to get a *mass* moving you have to push it. You have to apply a *force.*

You don't need to apply a force to *keep* a thing moving. It will keep moving by itself, or it would if it weren't for things like friction, wind resistance, the force of gravity, or bumping into other things. (After all, what do we mean by moving? Moving relative to *what?*) No, you only need apply a force if you want to *alter a body's motion* in some way. Or to *accelerate* it, as we say.

Newton said force equals mass times acceleration, or F = ma. (Remember we told you back on page 38 that you should become familiar with this equation.) For example if you want to *accelerate* twice the mass you have to apply *twice* the force. (You'll find more on this on page 169.)

* A pronoun, according to Webster, "used in a generic sense, *or when the sex of the person is unspecified.*"

When you apply a *force* to a *mass* to get it moving, the energy is converted into what Newton called *kinetic* energy (from the Greek *kinetikos*) or energy of motion. *Remember this is what they said in Newton's time*. According to scientists back then a body in motion possessed energy simply by virtue of the fact that it was *moving*.

And it's precisely because bodies in motion possess *energy* that people get killed in automobile accidents.

The (kinetic) energy possessed by a body is proportional to the *square* of its speed, scientists said, and this is correct *up to a point!* For example, a car traveling at sixty miles an hour has *four times* the energy of one traveling at only thirty miles per hour. (So think how much energy it would take, *even under this law*, to get something moving at the speed of light!)

Einstein wondered what would happen to a body if it were to approach the speed of light.

Protons, with neutrons and electrons, are basic parts of an atom, or *subatomic* particles. All atoms have protons and electrons, though they don't *necessarily* have to have neutrons. Hydrogen atoms don't have neutrons (page 78) and there's more hydrogen in the universe than all the rest of the elements put together.

Atoms themselves are tiny. If you wanted to see the atoms in a basket ball you'd have to make the basket ball as big as the earth. Then an atom would be about the size of a grape.

You'd have to make the grape-size atom as big as the Taj Mahal* before you could even *see* any of the sub-atomic particles.

But protons are one of the few things that we can accelerate to near the speed of light. They're much too small to push, but a force can be applied by exposing them to radiation.

If the proton is at rest its subsequent motion will be determined by the law we keep talking about, Newton's law of motion which says $F = ma$. That's easy enough to use. And as we're talking about acceleration here we can transpose this into:

$$a = F/m.$$

But what if the thing we were trying to accelerate was *already moving* with a velocity of v? If v was very great, approaching the speed of light, *then this law of Newton's might have to be modified!*

* We threw this in because at the time we were negotiating to get this book published in India!

Before we go any further let's pause for a moment and review where we are so far.

Below we show how the Lorentz Transformation applies to *time*, to *length* and to *mass*! All of these, *if the object is moving very fast relative to the observer*, will be seen differently than would have been the case had the observer been moving along with the object.

(a) Time in a *moving* frame of reference *as seen by a person not moving,* which we'll call t', is equal to time in the *stationary* frame t multiplied by the relativistic conversion factor. He sees time passing more slowly in the *moving* frame of reference.

$$t' = t\sqrt{1 - v^2/c^2} \quad . \qquad \text{Or } t = \frac{t'}{\sqrt{1 - v^2/c^2}} \quad .$$

(b) The length of a *moving* body, *seen by someone not moving,* which we'll call L', is reduced, making the object appear shorter in the direction of travel.

$$L' = L\sqrt{1 - v^2/c^2} \quad . \qquad \text{Or } L = \frac{L'}{\sqrt{1 - v^2/c^2}} \quad .$$

(c) Finally the mass m' of a *moving* object (its *inertial* mass, remember) can get very large, and will be equal to its *stationary* mass m *divided* by the relativistic conversion factor.

$$m' = \frac{m}{\sqrt{1 - v^2/c^2}} \quad . \qquad \text{Or } m = m'\sqrt{1 - v^2/c^2} \quad .$$

These are the simplified forms of the Lorentz Transformation that we will be using throughout this little book where appropriate.

Chapter Thirteen continues on the next page.

Now we've said that when you accelerate a *mass* by applying a *force* then the acceleration is given by Newton's law of $F = ma$. On page 84 we transposed this into $a = \frac{F}{m}$.

We also said that if the mass was moving near the speed of light then Newton's laws don't work very well.

Einstein says that because of this instead of $a = \frac{F}{m}$ we have to have a new law. The new law should be $a = \frac{F}{m}(1 - v^2/c^2)^{3/2}$, he says.

Now all we've got to do is find out where clever Albert got this from!

As none of the books at the library show how Einstein got this new equation we'll just have to try and figure it out for ourselves.

We've shown how *time*, *length (or distance)* and *mass* all have to be *transformed* for things moving near the speed of light.

Time and length both have to be *multiplied* by the relativistic conversion factor. For mass we have to *divide* by this conversion factor. (Inertial mass *increases* at high speeds, eventually nearing infinity.) But how about the *acceleration*? After all, we're trying to find out why Newton's equation has to be modified and Newton's equation includes *acceleration!* So what exactly *is* acceleration?

The best known acceleration is the acceleration due to gravity, or g (sometimes known as Newton's Constant). This is the rate at which your speed would be increasing all the way to the ground (if we ignore wind resistance) if you leaped off the top of a tall building. When people talk about g forces they mean how many times greater than g is the acceleration to which the body will be subjected. The value of g is 32·2 *feet per second per second.** Note that acceleration is measured in *distance,* divided by *time squared!* (Refer back to page 50 where this was explained.)

It might make it easier if we temporarily replace *acceleration* (a) with *distance* (d) divided by *time* (t) squared. (Again, this is something we warned you about back on page 50.)

Instead of $a = \frac{F}{m}$ we now have: $\frac{d}{t^2} = \frac{F}{m}$. Or $m = \frac{F \times t \times t}{d}$.

We can develop this a bit more on the next page.

* At least that is the acceleration due to gravity *at the surface of the earth!* Other places would be different! On the moon it's about one sixth of that. On Mars gravity is about 38% of that on earth.

Now we've already said that t, d (same as L) and m all have to be corrected for conditions near the speed of light. But what should we do about the force, F?

At first glance it wouldn't seem that F needs to be transformed. However, this force is being applied to a body which is *moving at very high speed*, and from a stationary frame of reference it will be seen as $F\sqrt{1-{}^{v^2}/_{c^2}}$.

This is because the *effect* of the force, that is to say the amount of *energy* imparted, depends not only on the *magnitude* of the force, but on the *time* over which the force is able to act. As seen from the *stationary* frame of reference the force acts for a shorter and shorter time the faster the body is moving.

As it approaches the speed of light the body hardly has time to get pushed at all! The force is thus *effectively* less.

Going back to the equation from the bottom of the previous page instead of $m = \frac{F \times t \times t}{d}$ we now have

$$\frac{m}{(1-{}^{v^2}/_{c^2})^{1/2}} = \frac{F\,(1-{}^{v^2}/_{c^2})^{1/2} \times t\,(1-{}^{v^2}/_{c^2})^{1/2} \times t\,(1-{}^{v^2}/_{c^2})^{1/2}}{d\,(1-{}^{v^2}/_{c^2})^{1/2}},$$

This can be transposed into:

$$ {}^{d}/_{t^2} = \frac{F\,(1-{}^{v^2}/_{c^2})^{1/2} \times (1-{}^{v^2}/_{c^2})^{1/2} \times (1-{}^{v^2}/_{c^2})^{1/2}}{m}.$$

And changing ${}^{d}/_{t^2}$ back to a we get

$$a = \frac{F}{m}\,(1-{}^{v^2}/_{c^2})^{3/2}$$

And this, of course, is the equation that Einstein came up with to replace Newton's simpler equation of $a = {}^{F}/_{m}$.

This equation is fundamental, because it shows that, no matter how great a force is applied, or how small the body, when the body approaches the speed of light the acceleration will be zero.

However hard you push it, once you get near the speed of light the body just won't go any faster.

So that takes care of *one* of the things that Einstein said had to be changed. Instead of $a = {}^{F}/_{m}$ we've shown how this new equation of Einstein's that replaces it can be derived from first principles.

But we're not through yet, because we *now* have to consider the question of how much *energy* a moving body possesses.

In Newton's time people said that the energy of a moving body (which was equal to the *work* done in accelerating it, they said) was equal to ½ mv^2. So let's start by trying to find out where they got this from.

It was believed that mass never changed, and had nothing to do with how fast the body might be traveling. The work done in pushing a body to accelerate it (as we just said) goes into increasing the *energy of motion* of the body. Work and energy of motion are the same thing then. (But only in Newton's time, remember.)

So, under the old rules, when we talk about *energy of motion* we can use the term W, the *work* done in accelerating it. (We just agreed these were the same thing!)

But how did they come up with his kinetic energy equation back in the seventeenth century?

To develop it we'll use the following terms:

W = the *work* we put in (= the *energy of motion* of the body).
v = the *speed* of the body at the end of its acceleration.
m = the *mass* of the body.*
F = the *force* used to get the body moving.
t = the *time* over which the force acts.
a = the *acceleration* of the body, due to the force applied.
d = the *distance* traveled by the body.

(Note. If a body starts from rest and accelerates *evenly* to a speed of v, then its *average* speed will be *half* this, or ½ v.)

We know that the work put in equals *force* times the *distance*, or $W = F \times d$. Also that $F = ma$ (or ${}^{F}\!/_{a} = m$) and that $d = {}^{v}\!/_{2} \times t$.

Also that $v = a \times t$, or $t = {}^{v}\!/_{a}$. These are things we can use.

* This assumes *mass* is a constant, which is why the law had to be modified more than two hundred years later when special relativity showed how a body's (*inertial*) mass can change.

By substituting we can take the following steps in order to derive the old equation for energy of motion, $W = \frac{1}{2}mv^2$.

Starting with Work = Force x Distance ($W = F \times d$) we'll attempt to show how they came up with this law of $W = \frac{1}{2}mv^2$.

$W = F \times d$	but $d = v/2 \times t$	therefore....
$W = F \times v/2 \times t$	but $v = a \times t$	therefore.....
$W = F \times at/2 \times t$		or...
$W = F \times a/2 \times t^2$	but $t = v/a$	therefore....
$W = F \times a/2 \times v^2/a^2$		or....
$W = \frac{1}{2}\left\{F/a\right\} \times v^2$	but $F/a = m$	So $\underline{W = \frac{1}{2}mv^2}$

This is fine. *However*, as we said on page 83, we already know that the old laws don't work well near the speed of light. Einstein says the equation used in Newton's time for energy of motion isn't any good either! (He just won't give poor old Sir Isaac a break, will he!) Einstein says instead of the old $W = \frac{1}{2}mv^2$ we have to use a new equation.

Albert doesn't say how he came up with this. He merely says that due to special relativity the energy of a body is no longer given by the equation of $W=\frac{1}{2}mc^2$ but by a new expression

$$E = \frac{mc^2}{\sqrt{1 - v^2/c^2}}$$

This is important! Notice Einstein calls the energy E, and not W, because under relativity the energy is not merely the work done in accelerating the body. It is the body's *total* energy, and includes the energy it had (its *mass*) when it wasn't moving.

The author has been unable to discover how this new equation of Einstein's was derived. Some books on the subject simply jump from the old equation to the new one without any attempt to explain how to get from the one to the other. But in this little book we try to explain things step by step as we go along.

So now we have to figure out exactly *how* Einstein got from the old law, which said that the energy of a body was ½ mv^2, to this new equation of Einstein's which says that

$$E = \frac{mc^2}{\sqrt{1 - \frac{v^2}{c^2}}}.$$

If we can do *that* the rest of this project is going to be a piece of cake!

Here's how we *think* Albert *might* have come up with his new equation. (You don't have to follow all this. If you want you can skip it and pick it up again at the beginning of chapter fourteen.)

The reason we're trying to show how the equation was derived is because it is from *this* that we get Einstein's famous equation of $E=mc^2$. Also we want you to know that everything's above board and we're not using smoke and mirrors!

The problem is that *energy, mass* and *the speed of light* all meant different things to Newton and to Einstein. So let's take a look at these things one at a time.

Much of this follows from chapter eleven, where we showed that energy and mass can be changed from the one into the other.

First let's look at mass. To Newton mass never changed. An object traveling at high speed had the same mass it had when it was standing still. Speed had nothing to do with a body's *mass.*

Einstein, however, took a somewhat different view. Mass, he said, was a measure of how much *energy* a body had. The body had a certain mass before we started to push it, then gained additional mass as it was being accelerated. Eventually, as the body nears the speed of light, its mass becomes so great that you *just can't make it go any faster* no matter how hard you push it!

To Newton, as we just said, there was only one mass which he called m. To Einstein a body might have a mass of m before we start to accelerate it, but this increases until, at the speed of light, the mass becomes *infinite*! (Refer back to page 71 on this.)

So how do we find the *mass* of a moving body? Well earlier (p. 85) we found out how to modify things for them to be correct at speeds close to the speed of light. We must apply (in this case divide by) the relativistic conversion factor.

So now instead of m we must use m', where $m' = \frac{m}{\sqrt{1 - v^2/c^2}}$.

(m' is *always* the body's *inertial* mass, and depends on its speed.)

How about the speed of light? In Newton's time people were only beginning to realize light takes time to travel. They knew it traveled very fast, but that was about all.

Even by the old laws, at the speed of light the kinetic energy of a body, that is its *energy of motion,* would be large, enormous in fact, because the speed of light was so great, and because the *energy of motion* (the only kind of energy it had) was proportional to the *square* of this speed. But to Newton, other than that it went very fast there was nothing special about the speed of light.

To Einstein, on the other hand, the speed of light was *very* special. It was the maximum speed limit in the universe (see page 71). And at the speed of light the old laws go out the window.

Finally of course we have *energy!* To Newton this was simple. When you apply *force* to a body, to *accelerate* it, the body goes faster, and *all the work* done pushing the body goes in to the *energy of motion.* (Remember, we said we'd call it *kinetic* energy.)

Even under the old laws the energy is proportional to the *square* of the speed. This means at high speeds the energy of motion can get very large. On the other hand, if the body is just sitting on the table and not going anywhere then it has *no energy at all!* So said Sir Isaac Newton.

Einstein didn't agree. He said when a body's sitting on a table and not going anywhere, even though it's not moving it has *a whole lot* of energy, in the form of its *mass.* He *did* agree that in *accelerating* a body you're doing *work,* thereby giving it *energy.* Only this energy shows up as an increase in the body's (inertial) mass, Albert said.

So just where does this leave us in trying to figure out where Einstein got his equation? Bear in mind that in going from the old laws, which work for speeds we normally encounter, to Einstein's laws, which take account of special relativity and work at *any* speed, we're still dealing with the familiar case of *transforming* something that's valid in a *stationary* frame of reference into something that's valid when we approach the speed of light.

But you'll point out that the old law of $W = \frac{1}{2} mv^2$ (which is the law that everyone was using until Einstein came along) deals with bodies *in motion* not bodies that are *stationary*. That may be so, but in the first place the speed is normally insignificant compared to the speed of light, so for all intents and purposes we're dealing with a stationary frame of reference. And secondly the equation is not quite accurate, *even at low speeds.*

The old equation (which says that the *kinetic* energy of a body equals $\frac{1}{2} mv^2$) contains mass m, and we've found out (page 85) that m is *not* a constant. So when the body is moving at speeds in the tens-of-thousands-of-miles-per-second range, and its (inertial) mass increases, then the old equation can no longer be used.

At normal speeds the inaccuracy is too small to be detected. Even so, *in reality the inaccuracy starts as soon as the body begins to move!*

So we have a paradoxical situation here, because the only time the old equation for energy of *motion* is <u>totally</u> accurate is when the body is *not moving at all!* We point this out so as to justify the assumption that we're dealing with a situation where conditions in a *stationary* frame of reference have to be *transformed* into conditions in a *moving* frame of reference.

First let's take a look at how Sir Isaac Newton would have dealt with a body traveling at the speed of light.

If energy of motion is $\frac{1}{2} mv^2$, as was believed in Newton's time, then if $v = c$, instead of $\frac{1}{2} mv^2$ we could write $\frac{1}{2} mc^2$. (It's the same thing.)

Today thanks to Einstein and Lorentz we know that when a body reaches the speed of light its mass becomes infinite. And if m equals infinity, *then according to this theory* $\frac{1}{2} mc^2$ must equal infinity too! (Anything multiplied by infinity must equal infinity.)

On page 88 in deriving the equation we said the *average* velocity was *half* the final velocity because it started out at zero and increased steadily until it reached its final velocity. But where $v = c$ then, according to special relativity, no amount of additional energy will increase the body's speed. And if the speed is not increasing then the *average* velocity and *final* velocity are *the same thing.* This means the *half* disappears, and instead of the average speed being $\frac{1}{2}$ v it becomes simply v.

Similarly, instead of the *energy of motion* being ½ mv^2 it now becomes mv^2. Check it out on page 88 & 89. (Remember this whole argument is based on the special case where $v = c$.)

If the body's *energy of motion* is now mv^2, then at the speed of light where $v = c$ the *energy of motion* will be mc^2. And at the speed of light the energy of motion becomes infinite!

But this is wrong! The value of mc^2 may be large, but it's not *infinite!* The trouble is we've forgotten what we said a moment ago. We've jumped from a stationary frame of reference into a moving frame of reference, and the only way to do *that* is by using the Lorentz relativistic conversion factor.

We have to replace m with m', where $m' = \dfrac{m}{\sqrt{1 - v^2/c^2}}$.

So a few moments ago, when we said (first) that the energy of a body instead of being ½ mv^2 at the speed of light was ½ mc^2, then (next) when we said that if v is not increasing the *half* disappears and the energy was now mc^2, what we *should* have done was replace m with m'.

So if $m' = \dfrac{m}{\sqrt{1 - v^2/c^2}}$ then the energy, instead of being m times c^2 would be $\dfrac{m}{\sqrt{1 - v^2/c^2}}$ times c^2. Or $\dfrac{m c^2}{\sqrt{1 - v^2/c^2}}$.

And this gives us Einstein's equation of $E = \dfrac{mc^2}{\sqrt{1 - v^2/c^2}}$, which replaces the old equation we all grew up with, which said that the kinetic energy of a body was ½ mv^2, and which was explained in detail on pages 88 & 89.

And at speeds much less than the speed of light both equations give virtually the same value for the body's *energy of motion.* The major difference being that Einstein's idea of total energy includes the energy possessed by the body due to its *stationary* mass (the *mass* it has when sitting on the table and not going anywhere) while to Newton *energy* and *mass* were two entirely different things.

CHAPTER FOURTEEN.

SHOWING THAT $E = mc^2$.

Now we've discussed how mass can be changed into energy, such as in nuclear weapons or nuclear powered aircraft carriers or submarines. We also mentioned the thing for which Einstein is most famous (his short three-page paper, remember) namely the equation $E = mc^2$. But we didn't offer any evidence, so here goes.

Einstein says the energy of a body E is given by $E = \frac{mc^2}{\sqrt{1 - v^2/c^2}}$.*

The previous chapter dealt with how we think this equation was developed. So now let us go on from there.

The equation gives us the <u>total</u> energy the body possesses. And remember, Einstein calls it E (the big E). But where did all this energy come from?

Some of it came from the work done in applying a *force*, over a *distance*, when we were pushing the body to *accelerate* it.

Now to Newton this work done in pushing the body to make it go faster just went to increasing the *kinetic energy* of the body. You couldn't measure this kinetic energy directly, and the only way you knew the moving body *had* any energy was because it could cause a lot of damage when it hit something.

But in chapter eleven we told you how Einstein determined that this energy *did in fact* show up in a form that could be measured. *This work went to increasing the body's (inertial) mass*, he said.

So now we knew that *some* of a moving body's mass derives from the *work* done in accelerating the body to its present speed.

But not *all* of the mass came from the *work* done in pushing it. The body already had *some* mass before we started pushing!

What we'd like to know now is, *how much of the total energy resulted from the work done in pushing it, and how much (in the form of its mass) was already there to start with?*

Read this last sentence again, it's important.

* See page 89, also appendix (6).

Let's examine more carefully this new equation of Einstein's which, he says, gives us the *total* energy of a body at *any* speed.

Sometimes a math. expression can be stated in another way. One way is to convert the expression into the form of a *series*.

Expressing Einstein's new equation $\dfrac{m\,c^2}{\sqrt{1-v^2/c^2}}$ *as a series* we get

$mc^2 + \frac{1}{2}mv^2 - (\frac{3}{8}m\frac{v^4}{c^2}) +$ ....and so on in diminishing sequence.

The above is *just a different way* of stating Einstein's equation for the *total* energy of a body. (Expressing things in the form of a series is not something Einstein invented and needs to be justified here. It's a device you'll learn about in college if you decide to study mathematics or engineering.)

Looking at the series two things stand out:

(1) The first step of the series, mc^2, does not contain v *and so has nothing to do with velocity!*

(2) The third step, and all subsequent steps of the series, due to their containing functions of c in the denominator, are small enough to be ignored under normal conditions where v is small compared to the speed of light.

This leaves us with the second step of the series, $\frac{1}{2}mv^2$. *This is the old value for the energy of motion, as used back in Newton's time, and is what engineers still use every day.* (We derived it back on pages 88 and 89, remember.) It's what we *always* use unless we are dealing with things moving near the speed of light! This shows that *for ninety-nine percent of the time the old laws are all we need.* (Of course if the body is moving at speeds near the speed of light then, *because the body's inertial mass has increased and m is no longer a constant*, the *whole* of the series has to be used.)

Let's go back to the first step of the series again, mc^2. We said it *did not contain v* and therefore had *nothing to do with velocity!*

Now the only part of the total energy package that doesn't have anything to do with velocity is the energy equivalent of the body's (stationary) *mass*, that is to say the energy it had before we started, when it was sitting on the table and not going anywhere. Before we started pushing it! When it *had* no velocity at all!

That's what we've been trying to find out! Because if we're right, and if mc^2 is the energy the body had before we started to move it, that is to say the *energy equivalent* of its *stationary mass,* then the rest of the energy must be what the body gained as a result of the *work* (W) we did in pushing (accelerating) it.

Now we said that $\frac{mc^2}{\sqrt{1 - v^2/c^2}}$ is all the energy possessed by a body, and, as we've already said, Einstein called it E, for *the total energy.*

So now, based on what we've just found out, we know that the total energy E, which is $\frac{mc^2}{\sqrt{1 - v^2/c^2}}$, must be made up of

(a) The energy the body originally had in the form of *mass* (mc^2) when it was standing still, and...

(b) The *work* (W) we did in pushing it!

Based on this we can say that $E = mc^2 + W$. And if we're *not* trying to make the body go faster, if we're *not* applying a force, over a distance, and we're *not* doing any *work*, then W = zero!

This means if the body's *not moving at all*, if it's just sitting on the table and not going anywhere, then where W=zero, now instead of saying $E = mc^2 + W$ we can simply say that $E = mc^2$.*

There you go!

To sum up, if a body is not moving (or even if it *is* moving, *but just so long as it isn't going anywhere near the speed of light*) the energy equivalent of the body's *mass* is given by that most famous equation of the twentieth century, $\underline{E = mc^2}$.

* Appendix (5) shows what a lot of *energy* can be obtained from very little *mass*.

We've shown how *mass* and *energy* can be changed from the one into the other. In a nuclear reactor, for instance, as well as in nuclear weapons, *tiny* amounts of mass are changed into *enormous* amounts of energy. Soon after the universe began (after about three hundred thousand of our current years, it is thought) for the next few million years that followed *energy* was being converted into *mass*, initially into elementary particles, then into atoms of hydrogen, the simplest and still by far the most abundant of all the elements.* We haven't had much success doing *that* yet (not outside of a laboratory anyway).

That Einstein had a tendency to complicate things he admitted many years later.

It was in 1944, during World War II, that plans were being laid for his original 1905 paper *Zur Elektrodynamik bewegter Korper* (on the electrodynamics of moving bodies) to be auctioned off with the proceeds to be used for the purchase of war bonds. The document was then to be placed in the Library of Congress. (One wonders why a document written by a foreign national, in a foreign country, and in a foreign language, should end up in the Library of Congress!)**

Unfortunately Einstein's original (1905) manuscript***had long since been thrown away, no-one back then having had any reason to suspect it might one day be of value.

* It's likely there was some helium present right from the start, and probably a small amount of lithium too.

** In his will Einstein left all of his memorabilia, not to the Library of Congress but to the State of Israel (which had meanwhile come into existence in 1948) though he chose not to reside there and even declined the offer of that nation's presidency. So his loyalties would seem to have been, at best, somewhat confused!

*** The word manuscript of course *means* hand written (as opposed to printed). It is a composite of *manus*, a fourth declension Latin noun meaning hand, and *scriptum*, meaning that which is written. Unfortunately hardly anyone speaks Latin any more, and while Einstein's 1905 papers in all likelihood *were* hand written, the word manuscript has since come to mean any original document, whether or not pen or pencil were employed in its preparation.

Einstein offered to produce a facsimile by copying from the particular edition of the *Annalen der Physik* in which the paper had originally appeared. This was done according to plan, the copy being sold for the sum of six and a half million dollars (to the Kansas City Insurance Company) and duly placed in the Library of Congress.

In preparing to make the copy Einstein asked his secretary, Helen Dukas, to read the original paper to him since he'd forgotten exactly how he'd presented the various arguments at the time.

Years later Ms. Dukas told how Einstein looked up and asked if that was actually what he'd written. Assured that it was he remarked that he realized now he could have said it more simply.

In 1921 Einstein won the Nobel Prize for physics, which back then was worth around $32,000 (a lot of money in those days).*

Albert may have got the glory, but his former wife Mileva got the money!

It being generally acknowledged that a Nobel prize was sooner or later inevitable, the proceeds of this anticipated windfall had been reluctantly pledged to the departing Mrs. Einstein (and her lawyer) as part of the divorce settlement.

* The Nobel Prize was in fact awarded in 1922, but retroactive to 1921. It was awarded, not for his work on special relativity, but for the whole of his contribution to theoretical physics, "*especially for his discovery of the law of the photoelectric effect*" (page 11). There had been doubt as to whether special relativity was something "*from which humankind had derived great use,*" (a specific requirement for award of the prize under the terms of Alfred Nobel's will).

A FEW FINAL NOTES.

The rest of this book (until you get to the appendices) doesn't really have much to do with special relativity, but it's a sort of follow on in that it tells how the universal speed limit (namely the speed of light) enables us, by looking out in space, *to look back in time.* This in turn helps us to gain a better understanding of the structure of the universe.

Knowledge of the speed of light also impacts any hopes we may have of one day emigrating to other planets, either within or beyond our solar system.

The establishment of colonies on other planets is a matter that everyone seems to take for granted. Television presentations on the subject, mostly narrated by pseudo-scientists of dubious academic credentials, are wildly optimistic. The whole thing is made to seem absurdly easy, yet the truth is, even within our own solar system the problems associated with inter planetary travel are staggering.

Billions of years of evolution have fitted us, together with all other known forms of life, to survive within a very limited set of conditions, namely those prevailing on planet earth. No conditions even remotely similar have ever been found to exist elsewhere.

Science fiction assures us we shall soon move on to colonize the cosmos. The following notes may cause the reader to develop a more realistic assessment of the situation.

Though he may one day be proven wrong, it is the author's belief that settlements of human beings on any other planet, *such that they are self sufficient and in no way dependent upon earth*, is just not going to happen! The way things are going we'll all be extinct long before we ever reach that stage.

What may later be achieved by other species that may possibly come after us (if we haven't totally destroyed the planet for them) is equally questionable for the same reason.

A few final notes, continued.

Throughout this paper we've insisted that the speed of light is (about) 186,000 miles per second. This has been learned from many experiments over the last three hundred years or so. Today it can be measured very accurately indeed. Nevertheless there are instances in the past where the speed of light *appears* to have been exceeded.

For example, scientists are trying to figure out what happened during the first few *trillionths of a trillionth* of a second after the big bang ocurred.

If the universe began as a singularity (see pages 41 & 42) and promptly blew apart, then for the first several billion years at least gravity must have been *slowing down* the rate of expansion. So knowing how fast the universe is expanding today, just think how much more quickly it must have been expanding immediately after the big bang!

Scientists believe the initial rate of expansion of the universe was so incredibly fast that it took only a trillionth of a second, that is to say 10^{-12} (or 0·000,000,000,001) of a second to go from *smaller than an atom* to the size of our present solar system! Or just one micro-second to reach the size of a galaxy. (One wonders how anyone can possibly know this?)

If the scientists were right it would mean that back then, whatever it was of which the universe at that time consisted, must have been traveling *trillions* of times faster than the speed of light.

Alternatively, the speed of light back then could have been a lot greater than it is today. Or perhaps it's simply that the fundamental and universal natural laws, with which we are all familiar today had not yet, at that time, been established.

Earlier we said that in about five billion years from now all life on earth will be destroyed, as the sun gets thousands of times hotter. There is, however, an equally catastrophic event that might occur even sooner. We might not have as much as five billion years.

Although the universe continues to expand, not all the galaxies are getting farther away from one another. Our Milky Way galaxy is one of a local group of galaxies which are not flying apart. There are thirty galaxies in this group, none of them more than three million light years away. Or less than a million "parsecs" as an astronomer would say (see page 107).

A few final notes, continued.

Of the major galaxies in the local group the one closest to us is Andromeda, which is about 2·2 million light years away. It is also a galaxy much larger than the Milky Way. It seems that Andromeda is approaching us (or we're approaching it) at a speed of about three hundred thousand miles per hour (a little over 83 miles per second).

Now this is pretty slow, as things go in the universe. Still, sooner or later there's going to be a collision.

Galaxies usually have upwards of one hundred billion stars. Can you imagine what's going to happen when our two galaxies hit?

Most of the universe is empty space (see page 24). Stars are so thinly spread out that if you thought of one star as being the size of an aspirin located in the city of San Luis Obispo, say, then the next nearest star would be the size of another aspirin located ninety miles away in Santa Barbara.* So that individual stars in the two colliding galaxies, being so thinly dispersed, in most cases won't actually hit each other. But the gravitational attraction between the monstrous galactic masses will cause them to tear each other apart with such violence that the event will be equally catastrophic! The two super-massive black holes, one of which is now thought to be at the center of every galaxy, will merge into a single even *more* massive one.**

Although at the present rate it would take some five billion years before the Milky Way and Andromeda collide, and that's just about the time when the sun will be starting to seriously misbehave, this speed will *increase* as the two galaxies draw closer together. As a result the collision will likely occur in only about *four billion* years. (Another estimate puts it at nearer *three* billion.) But either way it seems the end of the world is going to come sooner than we thought!

* If the distances are *that* big, and remembering there are more than 10^{22} stars, just imagine how vast space must be to accommodate all this! (Of course you can't imagine it. It's unimaginable.)

** The black hole at the center of a galaxy is said to always account for about half of one percent of the total galactic mass. However, this is not consistent with other data that says the black hole at the center of *our* galaxy is only three to four million times the mass of our sun (page 35). And although another source says this figure is actually *thirty* million, this would *still* make *our* black hole much less than half a percent the mass of the Milky Way.

A few final notes, continued.

And last but not least, let's talk a bit about gravity, because it's something that crops up from time to time in this book. In chapter four, for example, and again in appendix (1). Other places too.

Gravity, the force with which we are the most familiar, is by far the weakest of the four* forces that occur in nature. Consider the following:

Suppose you take a small magnet, the kind used to post things on refrigerator doors. If you place a paper clip on the table you should be able to pick it up (the paper clip) by holding the magnet just above it. As soon as the magnet gets near enough the paper clip will jump up off the table and attach itself to the magnet. The whole earth, with a mass of some 5·9 billion-trillion ($5{\cdot}9 \times 10^{21}$) tons, is trying to pull the paper clip down onto the table, yet something as small as this little magnet can overcome it!

So you can see that, by comparison to the other three forces that occur in nature (if we don't count dark energy, see footnote) gravity is easily the weakest of them all. (Though this would probably be of little comfort to the sky-diver whose parachute doesn't open.)

And while we're on the subject of gravity there are those who think that, especially at great distances, it may not simply follow an inverse-square rule (see middle of page 116). Which would mean that gravity's a lot more complicated than anyone thought!

The other three forces that occur in nature are the weak nuclear force, the strong nuclear force and the electro-magnetic force (the powerful repelling force mentioned back on page 50).

Next we're going to look at how we think the universe is going to end. Because for a long time no-one knew whether the universe was going to stop expanding, then start contracting until eventually it would all come together in a singularity, or whether it would simply keep expanding for ever! A lot more has been learned about this over the past few years, however, and that's what we're going to be talking about on the next fourteen pages.

* Dark energy (see pages 116-118) seems to be a new force that until recently we didn't know about. So perhaps we should be talking about the *five* forces in nature instead of four! However, this whole subject of dark energy, and dark matter, and how gravity might be involved even, is an area in which clearly there is much more to be learned.

How Is Everything Going To End?

This next part has to do with the ultimate fate of the universe.

Some of you may be familiar with the work of a well known poet in which he speculates as to whether the world will end in fire or in ice.* Perhaps this is taught in school.

Well we already know how the world is going to end, don't we! We covered this back on page 79 & 80, remember? As the sun nears the end of its life, and as all the hydrogen in its center gets used up, the sun will undergo a series of changes, the first of which will be an enormous expansion where the sun becomes so huge it will engulf Mercury and Venus, and possibly even earth itself. All life will end then, if there's any life still left by that time, as temperatures reach thousands of degrees.

The other possibility, a collision of our galaxy with one of its neighbors, is reviewed on page 103. Either way, though, it seems the world is going to end in fire.

But how about the *universe?* Suppose the poet referred to above had been talking about the *universe* instead of the world. Is the *universe* going to end in fire? Or in ice?

This was a matter that occupied astronomers for most of the twentieth century. Only now, with the new tools at our disposal, has the answer to this riddle begun to emerge. *We now think we know how the universe is going to end.*

At the time this author was born everyone thought the universe was static and wasn't ever going to get either any bigger or any smaller. After all the Bible said the universe was created in seven days (for reasons which remain unclear) and was going to stay that way until the day of judgment. There was even a very clever Irish Archbishop, by the name of Ussher,** who calculated exactly *when* the universe was created, based on bits of information he claimed to have found in the Bible. (The devil can cite scripture for his purpose, Shakespeare wrote.)***

* Robert Lee Frost (1874-1963).

** James Ussher (1581-1656).

*** Antonio. The Merchant of Venice, Act 1, Scene iii.

God created the universe starting at 9-o-clock on the morning of Monday, October the 26th, in the year 4004 B.C., the bishop said.

The job lasted six days, and on the seventh day, a Sunday, God rested.

Now even though Charles Darwin had blown some holes in this theory this didn't *necessarily* mean the universe wasn't static. After all everyone had always *believed* the universe to be static. Maps of the sky made centuries apart revealed no changes taking place.

Still a static universe did puzzle some people. For one thing by 1916 Einstein's theory of *general* relativity seemed to indicate the universe had to be either expanding or contracting. What it *couldn't* be, according to this theory, was *static!**

Yet Einstein himself felt so sure the universe was static he said if his theory meant it had to be changing then he must have made a mistake! So he massaged his calculations, and added what he called a cosmological constant.** This was a force equal in magnitude to gravity but *opposing* it, thereby making a static universe possible.

But this supposedly static universe would have been somewhat tenuous for, as Eddington*** pointed out, even if Einstein was right, and there *was* a cosmological constant, due to the two forces varying (one directly and the other inversely) as the *square* of the distances involved, there'd be an inherent instability (not unlike that which we describe on page 133). Equilibrium being maintained *only so long as nothing moved*, the slightest jiggle anywhere and the universe would immediately start to either expand or contract.

* Even *without* general relativity there were indications that the universe couldn't *be* either static *or* infinitely old. It had to do with the brightness of the sky.

** Here the word constant appears to be used less in its accepted mathematical meaning, and more in the sense that it was the reason why everything in the cosmos was "*constant*" and unchanging.

*** British scientist Sir Arthur Eddington (1882-1944).

Then came a startling piece of news.

By 1929 Edwin Hubble had discovered a *red shift* in the wavelengths of light coming to us from distant objects. We'll be covering this red shift in appendix (1). It shows that just about everything in the universe is rushing away from us at high speed. *And the farther away a body is, the faster it's moving.* (See below.)

Measuring *how far* the waves have *shifted* towards the red end of the spectrum tells us *the speed* at which the object is receding.

The distances to some of the closer stars had already been found by using parallax trigonometry. This involved taking a sighting of a star against a virtually unchanging background of more distant stars then, six months later (from a position 186 million miles away on the other side of the sun) taking *another* sighting. Because the change in angle was so small, however, this was a method that could only be used for the very nearest stars.

Then it was found that a certain type of star, called a Cepheid variable, whose brightness fluctuated, emitted light in proportion to the *rate* of this fluctuation. (The slower the rate of fluctuation the larger the star and the brighter it shines.) So by measuring a Cepheid star's brightness, and comparing it to that of another Cepheid that was fluctuating at the same rate, and that was close enough for its distance to be calculated by the parallax trigonometry method described above, we could tell how far away the first star was.

The red shift showed that the *speed* of a receding body is directly proportional to the body's *distance* from us. Dividing this *speed* (in kilometers per second) by its *distance* (in mega-parsecs)* gives us a constant. Astronomers call this the Hubble Constant, and it works for all things in the sky (well *almost* all things, see pages 102/103).

* Astronomers prefer to measure astronomical distances in parsecs (for *parallax second of arc*). This is how far a body would have to be away if angular measurements taken six months apart differed by one second of arc (one degree divided by thirty-six hundred). This is about 3·26 light years or just over thirty-*trillion* (30×10^{12}) kilometers.

The Hubble Constant meant that if you were to trace the paths of celestial bodies *backwards* you'd discover that they must have all started out from the same place, and at the same time. It seems that at some time in the distant past the universe appeared as a single point and exploded outwards. (We discussed this back on page 41-42.)

The Hubble Constant turns out to be about 70* (plus or minus 5). Interestingly the information used in the most recent calculations of the constant was gathered by the *Hubble* Space telescope (p. 110).

Then on pages 42 thru 44 we tried to show how just about every body in the universe is moving away from every other body *yet there is no center from which they are all moving away.* We said you should think of galaxies as being like spots on the surface of a giant balloon which is being blown up and getting bigger and bigger all the time, yet nowhere *on the surface* of the balloon is there a center.

If the universe was expanding then there was no need for a cosmological constant. Einstein had invented it in order to come up with what *he thought* was the right answer! (Engineering students know if you try *that* you usually get in trouble sooner or later.) It was the biggest mistake he made in his life, Einstein later admitted.

Now we knew the universe was expanding the question *then* was whether it was *finite* or *infinite.*** If the universe contained enough *mass* then gravity would slow the receding bodies down until finally they'd stop altogether. Then the universe would start to contract, slowly at first, then faster and faster until finally everything would end in *the big crunch* (the opposite of the big bang).

Presumably after the big crunch the universe would be reborn in a new big bang and the whole thing would start over (though there's no reason why this new universe should have to be anything like the one that just ended). This was the theory of the *finite* universe!

If there was to be a big crunch it meant that all the energy and all the mass in the universe was going to collapse into a singularity, an infinitely small point where temperatures would soar beyond our ability to imagine. A finite universe would end in fire, not ice.

* Another source, expressed in different units, gives the constant as 16 miles per second per million light-years (which works out to be a bit higher).

** What an *infinite* universe means is that its <u>*space*</u> component becomes infinite. Mass/energy, it is thought, is itself *finite.*

Many books were published on the subject of this supposedly finite universe. One such book this author remembers from a very long time ago was written by a Hungarian professor by the name of Opik. It was called *The Oscillating Universe*. Because, the learned professor said, it keeps expanding and contracting.

Alternatively, if the universe had insufficient mass then it would keep expanding for ever. Eventually the sky would become dark as all the stars and galaxies drifted infinitely far apart. All activity would cease, the stars would burn out and temperatures slowly fall to absolute zero. The universe would end, figuratively, in ice.

We said that whether the universe was going to end in fire or ice depended on *how much* matter there was in the universe. Attempts to calculate this were inconclusive, however.

Still most people believed the expansion of the universe *had to* eventually stop. This was supported by evidence of something that astronomers call dark matter. This is cosmic material that emits no discernable radiation. It is everywhere but we can't see it. Nobody knows what dark matter is, but while it has mass it does *not* seem to consist of atoms! It's a whole new kind of particle and, together with dark energy (p. 116) is now thought to account for as much as ninety-six percent of everything in the universe. (Astronomers had long known that mass must exist beyond that which could be seen.)

The probable existence of this stuff suggested there was enough mass now for gravity to bring the expansion to a halt and initiate the collapse of the universe. One day there was going to be a big crunch.

Everyone was happy that the matter had finally been resolved. Professor Opik no doubt became an instant celebrity and appeared on all the Sunday TV talk shows in Budapest.

Such was the prevailing view among astronomers. But recent developments have called this conclusion into question.

Huge new telescopes have now been built. Most astronomical telescopes these days are of the reflecting type invented by Sir Isaac Newton. Since manufacturing huge mirrors is very difficult many smaller mirrors are used, and coordinated somehow to produce the effect of a single mirror equal in area to all of them put together.

Two examples are the giant Keck telescopes located atop the extinct (we hope) volcano Mauna Kea (it means *white mountain*) on the big island of Hawaii. They are the largest optical telescopes ever built, each consisting of thirty-six separate mirrors the equivalent of a single enormous mirror ten meters in diameter.

The first of the Keck telescopes came into service in 1992, the second eight years later. Furthermore it is possible to coordinate the two instruments to simulate a single enormous mirror providing a capability surpassing that of any telescope hitherto available.

And while on the subject of telescopes, remember on page 108 we spoke of the Hubble space telescope. This is an instrument which is in earth orbit so able to make observations free from the distortion caused by the light having to travel through earth's atmosphere.

Now a bigger and better space telescope is planned, designated NGST (for Next Generation Space Telescope). This telescope will concentrate more on the infrared part of the spectrum.

Remember we told you (on page 107) how the light from distant galaxies is *shifted* towards the red (or longer wavelength) end of the spectrum.* This means that some of the radiation the Hubble space telescope would *otherwise* have been able to detect, now consists of waves *outside the range for which the device was designed.* This is where the proposed NGST comes in, for it will have the capability to detect radiation across a wider band than heretofore.

Now let's go back to these great new telescopes (like the Keck that we just told you about) that are being used to search for things *billions* of light years away. Some of these telescopes are designed to observe relatively large portions of the sky, say about the size of a full moon. Others are intended to concentrate on smaller areas and focus with greater definition once the location of a target is known.

These great telescopes are usually built on mountain tops where the air is thin and less of a problem. For while the light reaching them could have taken *ten billion years* or more to get there, it's only in the last *millisecond* or so that it starts to get distorted by the earth's atmosphere.

The super-computer is an indispensable tool in modern astronomy.

* Light is directed through a prism, where refraction causes it to be split into its constituent rainbow of colors, from short-wave violet light to the longer wave red. This display is the *spectrum*, in which certain dark lines appear. The red-shift was discovered by Edwin Hubble in 1929. For more on this go to appendix (1).

There are also much better computers today, thousands of times more powerful than those used in the Apollo moon landings of the late nineteen-sixties and early seventies.

The result of all this is that, not only are we now able to observe bodies billions of light years away, but we have the ability to analyze the massive amount of data gathered. And all this over just the last fifteen years or so. The result is that a new picture is beginning to emerge, which we're going to talk about in a minute.

Even though the universe is not static it does take a long time before any change is big enough to be noticed. (See page 106.)

Yet there's one exception. Our ancestors would occasionally notice a star where there hadn't been one just a few days earlier. The Romans called them novae, this being the plural of the Latin word *nova*, meaning new. Then, after a week or two, the strange new star would be gone!

The ancient Chinese also noticed these new lights in the sky and called them guest stars, because they didn't stay very long!*

Now we talked about novae back on pages 34-36, saying how they emit huge amounts of energy for a relatively short period.

A nova is an explosion of incredible violence that *sometimes* happens when a star comes to the end of its life. It has to be a *big* star for this to happen.

Our sun will never become a nova. It's too small for one thing. And there are other reasons. There are also *super*-novae, which are what we're going to be talking about here. (You only ever seem to hear about *super*-novae. Just like no-one ever talks about models any more, they're always *super*-models!)

* It has been suggested that the ancient Romans and Chinese might have been more intelligent than the average person of our present time (see second paragraph, page 125). At least until, in the case of the Romans, they became impaired by the lead (Latin *plumbum*) of which extensive use was made in their *plumbing* systems.

Although the average galaxy has upwards of *a hundred billion* stars individual galaxies produce supernovae only rarely, about once every hundred years or so. And in the Milky Way galaxy less often even than that! However, there are so many galaxies that *one* of them, *somewhere*, is sure to produce a supernova before long. The problem is we're mostly looking at places *billions* of light years away and the light, when it *does* get here, is too faint even to be seen.

What the Romans thought to be new stars must have been supernovae in relatively nearby galaxies. But while such events do occur in our own galactic neighborhood they're few and far between. In the whole of the twentieth century only about half a dozen or so were reported that were bright enough to be seen by the naked eye.

The most recent of such nearby supernovae at the time of writing was named 1987A, this designation presumably denoting the year of its appearing, and its alphabetic sequence among such events should there be others in that same year.

Most supernovae, even though they release more energy than *five-trillion-trillion* (5×10^{24}) World War II atom bombs, are so far away that the only way to find them is by pointing our telescopes at the sky and searching tiny sections of it one at a time.

Because the light from these distant supernovae is too dim to be seen it is allowed instead to fall upon a sort of modern version of an high speed photographic (ancient Greek for light-writing) plate until enough photons have been gathered for an image of a supernova to appear (always assuming there was one there to start with of course). This will usually take anywhere from five minutes to half an hour. Sometimes it can take all night. Then we give the image to the computer, which will compare it to earlier data and tell us if it sees anything that wasn't there the last time we looked. Once in a great while we come up lucky and find what we're looking for.

Back in the chapter on electromagnetic waves we said there were two types of supernova, the Type 1 and the Type 2. But within the Type 1 there is a sub-type known as the Type 1-a.

A Type 1-a supernova occurs when the more massive of a pair of twin stars feeds off the other *until it reaches a certain critical mass.* Then it explodes with such violence that for a brief time it can be as bright as all the rest of the stars in its galaxy put together!

As this type of supernova occur *only upon reaching this same critical mass* (which is about one and a half times the mass of our sun) that means they all emit *the same amount of energy!*

So now we have a type of celestial body that is bright enough to be *detected* from more than ten *billion* light years away. And since they all emit *the same amount of energy* we can tell by their *apparent* brightness how far away one of them is compared to another.

The consistency of the energy emitted by Type 1-a supernovae means that we have a new yardstick (or "*standard candle*") for the measuring of great distances. This is because some of them are to be found in the same galaxies as Cepheid variables, stars whose *actual* distances can be calculated (page 107). Cepheid variable stars take their name from the constellation Cepheus, which can be seen in the northern hemisphere between Cygnus and Polaris (Cepheus being the constellation in which these variable stars were first discovered).

Incidentally a Type 2 supernova explosion is so violent that *in only a few minutes* it can radiate *a hundred times* more energy than our sun will emit during its *entire life of more than ten billion years.*

Back to the Type 1-a supernovae! Whether or not we've found one will be revealed by spectrographic analysis of the light. The presence of certain elements and the proportions in which they occur will identify it. (Spectrographic analysis of light is nothing new, it goes back to the middle of the nineteenth century. See page 110.)

The next step is to *record the position* of this light and view it through a telescope better suited to zero in on this tiny dot. Now we can measure how bright the supernova is (to tell how far away it is) and by how much its light is red shifted (or how fast it's going).

But how to find these supernovae? And where to look? Even an area of the sky so small it could be covered by a small coin held at arms length contains *hundreds of thousands* of galaxies, yet it could be days, or weeks, before a single supernova occurs *anywhere!*

However, using new wide-angle astronomical cameras, working with the giant Keck telescopes in Hawaii, and using supercomputers in the Lawrence Berkeley Lab. in California, a method was devised to *simultaneously* monitor large areas of sky, covering huge numbers of galaxies, so that eventually a small group of dedicated scientists were able to compile data on some forty-two suitable supernovae.

This was less than the fifty they'd been hoping for. But after several years of commuting back and forth to Hawaii the sun-tanned frequent flyers were starting to hear rumblings of discontent from the people responsible for paying the bills. So they put away their surf boards and reluctantly decided that the forty-two supernovae on which they'd collected data so far were just going to have to do.

The fact that we were now able to gather data on objects at such great distances provided us with a unique opportunity. Analysis of the data should reveal that the rate of expansion of the universe was continuing to slow down, *and if so by how much.* (This was what astronomers had been wanting to know ever since they first realized there'd been a big-bang.) The reasoning went as follows:

When the big bang occurred the universe began to expand at a prodigious rate (p.102). Almost immediately the rate of expansion began to slow down due to the effects of gravity.

Now when we look at something that is ten billion light years away we're seeing it as it was ten billion years ago! *And ten billion years ago everything was expanding a lot faster than it is today.*

In appendix (1) we will tell how the speed of a body can be found by measuring the red shift of the light coming from it. (You'll recall we've mentioned the red shift once or twice already.)

The trouble is that when observing radiation from such distant bodies what we're seeing is *gamma ray* emission, not visible light.

Gamma rays are of much higher frequency and nowhere near the wavelength of visible light as identified on page 22. *And there's no such thing as a red shift in gamma ray emission.*

Fortunately the gamma rays affect other nearby material, mostly clouds of gas and dust, causing them to glow and emit radiation in a wavelength in which the red shift *can* be measured.

So what we do is measure the red shift in the light from nearby material instead of from the gamma ray source itself. But it amounts to the same thing, because the material in question is known to be traveling at close to the same speed as the object of our interest.

Now back to what we were saying. If we measure the speed of something we see *as it was ten billion years ago,* then it should be traveling a lot faster than something we observe *as it was only five billion years ago.* (Because during the intervening five billion years everything should have been slowing down, right?)

This is something we have to take into account when considering the relationship between a body's speed, and its distance from us!

So what we do is study the forty-two Type 1-a supernovae that we just told you about. We establish their (comparative) *distances* from us by measuring their brightness, and we find their *velocities* from the red shift in light being emitted by nearby material known to be traveling at about the same speed!

Now we had forty-two *exactly similar* supernovae, all at varying distances from us, the farthermost of them being almost to the *edge* of the known universe.*

We were now able to examine, and with greater accuracy than ever before, the precise *relationship* between a supernova's *distance* from us, and the *speed* at which it was receding. This investigation was called SNAP, which stands for Super Nova Acceleration Probe.

Then came the biggest surprise since someone discovered the earth wasn't flat! Reviewing the various supernovae distances from us, consequently how far back in time we were looking, and comparing these to the speeds at which each of the forty-two sample supernovae were seen to be receding, the startling conclusion was reached that, not only was the expansion of the universe not slowing down at all, *but it was in fact speeding up!*

This unexpected result upset the theory that astronomers had held ever since the days of Edwin Hubble, namely that the universe began to expand violently at the big bang, *and ever since then the rate of expansion had been gradually slowing down.*

This new data showed that the expansion of the universe wasn't slowing down the way it should've been. In fact it wasn't slowing down *at all!* The rate of expansion, it seemed, after slowing down for billions of years, was now starting to speed up again!

This caused a sensation. Astronomers threw themselves from tall buildings! Television newsmen covering the event wept openly, while mobs of angry graduate students roamed the streets, smashing windows, overturning police cars and breaking into bookstores and libraries.

Everyone was talking about it!

This unexpected turn of events raised new questions. If the rate of expansion of the universe was *increasing* it meant that *earlier on*, for a while at least, it must have been expanding *more slowly.*

* We have to be careful when we talk about the *edge* of the universe since that suggests that something (space, for example) might exist *outside* of it, which is not the case. It all has to do with how the big-bang occurred *everywhere* simultaneously (pages 42 and 129).

Now if at one time the universe was expanding more slowly then it must have taken longer for things to get as far apart as they are today. So the universe must be older than previously thought.

If the universe was older than we'd thought then this explained one thing that had been puzzling astronomers for some time. There were stars whose ages had been reliably determined to be greater than that of the universe itself, which was not possible. (You can't be older than your mother.) But if the universe was older than we'd thought then this particular riddle was solved.

But why was the rate of expansion of the universe increasing? It was starting to look like there must be a strange new force out there, *a force which was pushing things apart.*

It was thought this new force must be the opposite of gravity. When bodies are close together this repelling force is too weak to be detected, it seems. But as bodies get farther apart from one another this force starts to get stronger. Astronomers have coined a name for this new phenomenon. They are calling it dark energy.

No-one knows what this dark energy is, but it's thought, together with dark *matter* (page 109) to account for as much as 96% of *everything in the universe.* It's the largest component of what is now being called the "standard model" where ordinary matter, consisting of atoms, accounts for perhaps only about four percent of the grand total. Dark matter is thought to make up another twenty-one percent, while the remaining seventy-five per cent or so is dark energy.

An alternative theory, a minority view, holds that gravity may no longer follow an inverse square law when the distances between the bodies are very great. As we said on page 104, this would mean that gravity is a lot more complicated than anyone thought it was.

So it would appear there *is* some kind of a repelling force after all. Einstein doesn't really deserve credit for this as he originally came up with the idea in order to solve a quite different problem.

In any case it's not the same thing. Dark energy doesn't simply keep the universe static like the cosmological constant was supposed to do. Because we now know that the universe is *not* static!

This new force evidently keeps getting stronger and stronger as the galaxies get farther and farther apart. Now it seems that dark energy has become strong enough that the rate of expansion of the universe, after slowing down for billions of years, is now starting to speed up again! And unless there's something else we don't know about (always a possibility) the fate of the universe is sealed.

Meanwhile, as we said, no-one knows what dark energy is. One theory has to do with the near-perfect vacuum previously thought to exist between galaxies (see page 24). The idea is that within this supposed vacuum tiny particles are continuously coming into, and going out of, existence, and as part of this process energy is being introduced. In fact one proposal is that instead of calling it dark energy a better name would be vacuum energy.

Whether or not this new concept means that the total amount of mass and energy in the universe could be changing after all is not clear. (See footnote on page 108.) The reader will recall that back on page 41 mention was made of the English astronomer Fred Hoyle who suggested that new matter was being created all the time.

But in any case this dark energy seems to have little effect until the space in which the activity is occurring is large enough. Then the energy of this increasing number of particles becomes significant, opposing gravity, *until the expansion of the universe is no longer slowing down as much as it had been.*

As the galaxies get farther and farther apart gravity has less and less effect while dark energy grows ever stronger. Now it seems the latter is starting to gain the upper hand! As gravity loosens its grip, and as dark energy continues to increase in strength, the fate of the universe is sealed! Not only will it continue to expand (remember, the *first differential* of distance with respect to time) but the rate of expansion (the *second* differential) and finally *even the accelerating force itself* will increase as the dark energy continues to grow ever stronger.

Shakespeare

The result is that the universe is destined to expand faster and faster to the last syllable of recorded time (as William Shakespeare's murderous monarch so eloquently put it).*

* Macbeth, Act V, Scene v.

And when we talk about the end of time we're not just using a figure of speech. For eventually time, that fourth dimension (see page 131) will have *ceased to exist.* Or at least it will no longer have any meaning.

Still if there *does* prove to be such a thing as dark energy then Einstein should at least be given *some* credit as the first person (for whatever reason) to have suggested its existence. To account for a state of affairs that turned out not to exist, the peripatetic professor came up with the solution to a problem that had not yet arisen!

What all this means is that our Hungarian professor was wrong and there's not going to be a big crunch. Apparently the universe is like that circle in the water that Shakespeare talked about (page 19) which, he said, *"never ceaseth to enlarge itself."*

An incredibly great (by our standards) yet *finite* quantity (the universe) by never ceasing to enlarge itself, so endlessly diluted by broad spreading into an ultimately *infinite* space must, in the end disperse to naught. (Remember that footnote back on page 108.)

So the end of the universe (if it can truly be said to end at all) is apparently going to come not with a bang but with a shiver. And while there won't be any ice (because ice has *mass*, and all mass will have dissipated into the infinity of space) what there *will* be is a total absence of *anything.* An endless void at a temperature of *zero* degrees absolute! So cold that molecules (if there were any) would have no motion, and gas (if there were any) would have no volume.

The wheel will have come a full circle. It will be as if there'd never been a universe at all, *that nothing had ever existed!* Time will have no meaning! There'll be no beginning and no end, no past, no present and no future. Matter itself will slowly decay, eventually leaving only an incredibly thin haze of elementary particles floating in an emptiness beyond our ability to conceive.

And the rest, as the dying Prince of Denmark said, is silence.*

I find this all very
discouraging.

* Hamlet, Act V, Scene ii.

What's the story on all these Flying Saucers we keep hearing about?

Are we alone in the universe? We and our fellow creatures, the dogs and the cats, the algae and the bacteria, and all the other living things. Is this the only place where life exists, or does it exist in many other places too, as most astronomers seem to think? (See p.173.)

Walking on the beach on a clear starry night, if you picked up a handful of sand, the number of stars you'd be able to see in the sky is fewer than the number of grains of sand you'd be holding in your hand. Yet the *total* number of *all* the stars is greater than all the grains of sand on all the beaches (and all the deserts) of the world.

So surely, the experts say, there must be lots of other planets like ours out there. In fact, they say, there must be millions, even *billions*.

But if there *are* countless other planets like ours out there, and if advanced civilizations have developed on at least *some* of them, then why haven't we heard from them?

Do all civilizations, upon reaching the level of technology where the transmission of such signals would be possible, promptly destroy themselves and their environment (like we're doing)? Has the great architect of the universe imposed a limit on how high a civilization will be permitted to build its technological Tower of Babel?

Unlike most people this author is not *necessarily* convinced that there is, or ever has been, life anywhere else in the universe. *There's no evidence of it anywhere.* Not even any radio signals from outer space (more on this in a moment). Such bodies as we've been able to explore so far have shown no signs of life, neither past nor present (despite a tendency to tilt the evidence in the direction of supporting such a conclusion). Everywhere we look conditions are found to be so extreme that life, as we know it, could not possibly exist. Such hostile environments appear to be the norm, and our own earthly Garden of Eden a magnificent exception.

And why *shouldn't* life be unique to planet earth? After all, no-one knows why the universe was created in the first place, or why any of us are here!

Einstein thought life was an illusion (a very persistent one, he admitted). Macbeth may have been closer when he said life was "*a tale told by an idiot, full of sound and fury, signifying nothing.*"

But let's get back to those signals from other civilizations that we talked about? Or rather the *absence* of any such signals.

We ourselves have been transmitting radio messages for about seventy years now,* long enough for the first of such messages to have reached a *mere* few *thousand* other star systems, though all of them in just a tiny corner of our own galaxy.

When a radio station transmits a signal to our homes the signal also spreads out into space. The sphere within which such a signal could *possibly* be detected is like an ever growing bubble, with the earth at the center, the bubble spreading out in all directions at the speed of light.

By now, if the signals we sent out back then were strong enough, there are a small number of star systems which, *if* they had planets, and *if* any of these planets had life, and *if* the life on any of these planets was of sufficient intelligence and had reached a high enough level of development, could *just possibly* be receiving our signals.

So our signals could, by now, *just possibly* have reached a mere *few thousand* other star systems out of the *billions of trillions* that are known to exist!** So to *almost all* of the universe our little planet earth would be a *silent* planet. If aliens searched the sky with their radio telescopes the overwhelming probability is *they'd find no trace of us at all!* There'd simply not have been enough time for even the *earliest* of our radio transmissions to have reached them yet!

But suppose some creatures on a planet of a nearby star system where our first radio signals were *just now* starting to arrive, suppose they detected a signal sent out from here on earth seventy years ago. This means the planet would be about seventy *light years* away.

Now this may seem like a long way away, but in the vastness of space such a planet would be, figuratively speaking, in our own back yard. Even so, our signal would have taken seventy years to get to that particular planet, and any reply the aliens might wish to send would take *another* seventy years to travel all the way back to us.

* Probably the first signal of sufficient strength, and a high enough frequency to punch through the ionosphere, would have been the short-wave broadcast of the Olympic games from Berlin in 1936.

** This is like covering the area of a postage stamp as compared to the entire surface of the earth!

That means if anyone wanted to ask these aliens a question, by the time the answer came back a hundred and forty years would have elapsed, and it would be the original questioner's great-great-great-great-grand children (or their contemporaries) who'd hear the reply. And this is a planet of a star system that's not far away, remember!

For over forty years now there has existed a program called SETI, which stands for the Search for Extra Terrestrial Intelligence. This program was originally funded by the U.S. government, but on Oct. 12th, 1992 (five hundred years to the day after Columbus first set foot in the New World)* this funding was cut, and the following year eliminated altogether. Fortunately the program was continued with money from non-governmental sources.

From time to time radio signals have been sent out by SETI in the hope of provoking a response. A few *possible* responses have been detected, one in 1987 reportedly featuring the characters 6EQUJ5.

An earlier possibility, picked up on August 15th 1977, lasted for 72 seconds and became known as the "wow" signal, after the remark scribbled in the margin of a computer print out by an excited SETI employee. But by the time additional resources could be deployed to investigate further the mysterious signal had disappeared, and to this day no other such promising signal has ever been detected.

Meanwhile if anyone knows what 6EQUJ5 means the people at SETI would like to hear from you.

The reason why communicating with planets of other stars is so difficult is because the distances involved are so enormous, almost beyond imagination. And because the strength of an electromagnetic signal diminishes as the *square* of the distance from its source.

In the past, space probes have been dispatched to distant parts of our own solar system, so far away that the radio signals they sent back took several hours to reach us. The signals when they *did* get here were weak, less than a *trillionth*, and by some accounts less than a *quadrillionth*, of a watt. (And a watt itself is very small.)

Now it's probable that if there *are* any aliens out there they're going to be living on planets of other stars, rather than in our own solar system. If so then any signals we pick up would have to have come from much farther away.

* For Columbus to have discovered America was remarkable. To have sailed west from Spain and missed it altogether would, as Mark Twain pointed out, have been *more* remarkable.

Consider again the planet that we talked about, the one that was part of a star system seventy light years away. This imaginary planet, even though it's very close (as star systems go) is some *hundred and fifty thousand* times farther away from us than were the space probes that we just mentioned.

We said the strength of a radio signal is *inversely* proportional to the *square* of the distance from its source. So any signal from our imaginary planet, by the time it got here would be more than *twenty-billion* times weaker than the signals from the space probes. It would have a strength of less than a *twentieth of a billionth of a trillionth* of a watt! (Or a thousand times less then *this* even, see previous page.) And while creatures from other civilizations could no doubt transmit much stronger signals than the ones transmitted by the space probes, especially since they'd probably have to be a lot more advanced than we are, the fact remains that any electromagnetic energy that could conceivably be projected, without destroying the body it was coming from, would have to be unbelievably weak by the time it had spread out seventy light years away from its source in every direction.

Remember these are our cosmic next door neighbors, on a planet that's not far away! Most stars, and any planets they might have, are located from *hundreds* to, in the case of the farthermost of them, *tens of millions* of times farther away from us than this one!

Now it used to be said the *total energy* of *all* the radio waves that had *ever* been received, by *all* the radio telescopes that had *ever* been built would amount to less than the energy of a single snow flake upon hitting the ground. (With all the radio telescopes there are today this may no longer be true.)

In any event we can pick up some pretty weak signals! But could we detect signals *sent by aliens* seventy or more light years away?

So the fact that we've not picked up any radio signals from other civilizations doesn't *necessarily* mean that there *aren't* any other civilizations. There may be, we just don't know!

But despite the difficulties described above, if we *should* ever be contacted by aliens from beyond the solar system it will almost certainly have to be by means of radio. For while this method of communication may have its problems, unless some amazing new developments take place in the years ahead (and providing there's still someone around to listen) then radio, or some similar form of electromagnetic radiation, is the only way we're ever likely to establish contact across the incredible vastness of interstellar space.

This is because the problems in communicating by radio are trifling compared to the difficulties anyone would face in attempting to physically transport intelligent beings, or robots, or *anything* for that matter, across the unfathomable emptiness between the stars.

The idea that we ourselves might travel to other star-systems is strictly science fiction. Other bodies *in the solar system* maybe. A landing on the planet Mars, for example, seems possible within a few decades if we set our minds to it (assuming our complex and fragile house-of-cards infrastructure, upon which such an ambitious project would depend, doesn't collapse before then).

It is a strange paradox that we human beings, while individually intelligent, *collectively* are remarkably stupid and shortsighted, a condition recognized and exploited by all successful politicians and destined to bring about our early extinction as species. For today mankind is facing a catastrophe we seem determined to ignore.

The population of our planet is exploding out of control, causing the rapid depletion of the finite resources upon which our continued existence depends. Furthermore mankind's profligate use of these resources is raising pollution to a level from which we may not be able to recover. The government and people of China, to their everlasting credit, are the only ones doing *anything at all* to avert the calamity that's staring humanity in the face.

In order to meet the demands of this ever increasing number of people we are plundering and polluting the fragile environment with scant regard for the consequences. Greedily we deny to other forms of life their rightful share of nature's bounty. Recklessly is nature exploited for man's exclusive benefit, in the process transforming our incredibly beautiful, and quite possibly unique, planet into a toxic wasteland where ultimately nothing will survive.

Now, after countless thousands of years of delicate balance, we are *each day* pouring into the atmosphere *tens of millions of tons* of greenhouse gases, trapping heat and causing the polar ice caps to melt, thereby reducing earth's ability to reflect heat, causing further temperature increases and so on. *Soon we shall have reached the point of no return!* Then, long before we've had a chance to find any other place to live, this earth, this realm, this precious stone set in a silver cosmos,* will be irreversibly transformed into a Venusian cauldron of poisonous gas at a temperature hot enough to melt lead.

* Shakespeare. King Richard II. Act II, Scene i. (Paraphrasing.)

Driven by the insatiable demands of an exponentially exploding population we are forced to insult, abuse and generally destroy what could well be the only body in the universe capable of supporting life. Certainly the only one we know about!

All of our environmental difficulties, indeed virtually all of the problems that beset mankind today, can be attributed either directly or indirectly to man's collective insanity in demanding the absolute right to procreate without restraint or accountability.

Over the lifetimes of many still alive today the number of human beings straining the resources of this groaning planet has swollen by a factor of roughly four! *By such irresponsibility is our expulsion from the Garden of Eden assured.*

There are countries today where sixty percent of the people are below the age of twenty! For one Pacific island nation the figure is said to be even higher. The burgeoning population of Latin America, a problem no doubt exacerbated by the outdated teachings of the church, exceeds the number for which those nations' resources are sufficient to provide. The resultant poverty is being exported to other countries, particularly the United States where failure to secure her borders has contributed to a noticeable decline in the standard of living, and to the establishment of foreign enclaves within her midst.

Yet foolishly, and in the face of all this, we applaud the parents of large families. On a recent television show the proud progenitors of no fewer than *sixteen* children were given a standing ovation for their achievement. *If this is not insanity what is?*

Throughout the ages the human race has endured hardship and deprivation. Now our ability to produce abundance is destroying us! In opposing nature's attempts to correct an evolutionary malfunction we only postpone, and worsen, the terrible day of reckoning.

Is the eradication of hunger and disease in developing countries such a good idea if the only result is a *further* increase in the number of people, and the consequent extending of the misery to everyone? *Absent a commitment to reduce the population*, should we *really* be bending every effort to make the situation *even worse?*

Selfishly we ignore the writing on the wall,* knowing that as individuals we'll all be dead before then anyway! Like governments busily incurring massive debt for our children to repay, we are careless of the interest of those who shall come after us.

* Daniel chapter 5, verses 24 thru 28.

Easter Island is a small island, and whoever it was that chopped down the last tree there had to have known what he was doing. It had been a land endowed with everything necessary for life to flourish. Yet by the islanders' failure to limit their population the formerly benevolent environment was destroyed. With the trees all gone the rain would wash away the fertile soil. There'd be no wood for boats from which to fish, no timber the few survivors could use to build a shelter. No fuel for fires to warm their emaciated bodies, or cook the flesh of the dead comrades that in desperation they were forced to eat. Only brooding statues, pleading to the gods for help that never came. It was a tragedy from which we should have learned a lesson. Yet now we're doing the same thing, only this time on a *global* scale!

In our arrogance we are rewriting the laws of nature, and will do so at our peril. Survival of the fittest is no longer the rule, and this must cause the genetic degradation of our species. Remember that nature doesn't give a hoot for what man may consider to be *ethical*. Concern for the weak and poor is contrary to her purpose. Evolution requires the weak to be *eliminated!* It places us in a quandary. For if nature is the manifestation of God's wishes, then it seems that organized religion must have somehow got it all wrong.

Today earth is home to more than six and a half *billion* people! *No-one seems to understand that this is the one horrendous problem from which all other problems spring.* While treating the symptoms we refuse to do anything about the disease. *Now the most drastic measures yet proposed would not even begin to solve the problem.*

Future generations may look back on this as an age of incredible stupidity on the part of their ancestors. Or perhaps the long overdue dawning of an awareness. But if we don't do something soon there'll be no-one left to look back on *anything!*

If we don't reduce our numbers then nature will do it for us! Perhaps we'll be ravaged by a disease for which we'll find no cure. Or maybe the end will come in a cosmic collision like those which, in the past, brought about the extinction of other malevolent or de-stabilizing species that had become a threat to the grand design.

And then again the destiny that shapes our ends may yet surprise us all. For chaos theory suggests that, unless we annihilate ourselves in a nuclear holocaust (the likelihood of which increases daily) what-ever it is that causes mankind to become extinct will probably be an event the possible occurrence of which had not been foreseen, and thus for which we were unprepared.

In view of the above it's unlikely that we're going to be around for long enough to develop the level of technology it would take for man (plus all the things he'd need for such a trip) to ever hope to cross the grand immensity of time and space to *any* other star.

But even should we manage to survive that long our launching pads of yesterday will stand abandoned and forlorn. By then there'll be more urgent matters demanding our attention here at home!

For if we don't do something soon we'll be experiencing famine and disease more terrible than anything the world has ever known. Crime will explode as civilization collapses. Competition for earth's dwindling resources will bring a level of violence that will make the recent episodes of genocide and mutilation in countries like Rwanda seem like bucolic interludes of tranquility by comparison!

As for sightings of UFOs (unidentified flying objects) and tales of little green men rescued from crashed space-ships, of government lies and cover-ups and stories of alien abductions, you won't find a single reputable astronomer who believes in any of this nonsense. In the almost sixty years since the first reported alien landing there's not been a single shred of evidence to support the notion that we're being visited by creatures from another world. This is perhaps fortunate, for if aliens from (what would have to be) an advanced civilization were ever to invade our planet they might treat us the way we treat those less advanced than *we* are. They might kill us for sport, or raise us under conditions of great cruelty for use as food. Or subject us to medical experiments while alive and conscious.

It is for this reason that some are suggesting it might be prudent to hold off from responding right away should any signal be detected coming to us from outer space. The prominent British astrophysicist and author Stephen Hawking (*A Brief History of Time*) agreed. We should keep our head low, was his (figurative) advice.

And why should aliens always have to look like *men*, anyway? Life evolves according to the environment in which it finds itself, and conditions on any other body where life might possibly exist would almost certainly be unlike anything we know. Denizens of such far pavilions* might not even be recognizable as life at all, let alone are they likely to look like *men,* green or otherwise.

* A romantic metaphor for distant lands (from the M.M. Kaye novel of that name).

As far as this former engineer is concerned, if there's no-one out there capable of even sending us *radio messages* then it's not likely that aliens are going to be arriving here any time soon. You're not going to wake up one morning and find a bunch of funny looking men with large heads climbing down from a flying saucer parked in your back yard. That so many should believe this to be possible is a sad commentary on the average person's level of understanding of the structure and enormity of the universe.

Unfortunately television, and the internet, potentially such great educational tools, which could have been employed to stimulate the nation's slumbering intellect, have instead been dumbed down to the intelligence level of the proletariat and used to stimulate the witless consumer's desire for things he can't afford and doesn't need.

So rather than sit all day in front of cathode ray tubes or liquid crystal displays, a visit to the library is recommended. Available there is all the accumulated knowledge of the countless generations who've gone before us. It usually doesn't cost anything to join, and you'll be surprised how much fun you can have!

Meanwhile, in bringing this book to a close, the author hopes his humble effort may have stirred within the reader a renewed interest in the mysteries of the cosmos and the significance (if any) of our place within it.

Free your mind to soar above the trivia of everyday events, to contemplate instead, with thought beyond the reaches of your soul, the wonder and complexity of that vast continuum of time and space being gradually revealed and of which, for just the blinking of a cosmological eye, we strangely find ourselves a part. To struggle, in the little time we have, towards the eternal truth. To ponder when, and how, and why.

For as the crazy Hamlet said, "*There are more things in heaven and earth, Charlie Brown.....*"

Appendix (1).

It should be noted that the universe is not really expanding out into space. The universe already fills all of space, and always has.

We tend to think there's unlimited space out there, and that the universe is continuously expanding out into it. But it's really *space itself* that's expanding, and has been ever since the big-bang. Space, like matter, energy and time, only came into being at the instant of the big bang!

Think of space as being *part of* the universe. It's what exists between the galaxies and is the part that's expanding.

Virtually all the galaxies are moving away from us. Generally the farther away a galaxy is the faster it's receding.

The *singularity* (the big bang, or the creation, or whatever you want to call it) was the beginning of everything! It introduced into the emptiness something that *became* the universe we know today.

One theory says it was energy that was introduced, and that it was this energy that began to turn into mass (the stuff that everything's made of). A variation says that what appeared was a sort of plasma, and that it was this plasma that then, after a while, started to change into the energy and mass that we know today. All we really know for sure is that for our universe the big bang was the beginning of *everything!* Time, space *and everything else* were only created at the instant of the big-bang, and *did not exist* before it!

So if space has not *always* existed, if it was *initially* contained within the singularity, then *the big-bang must have occurred everywhere simultaneously!* There's no one place where it happened.

Whatever way the universe may have started we do know that it began to expand, explode might be a better word, at an incredible rate (page 102). But because gravity was trying to pull everything together (the way it pulled the apple to the ground where Sir Isaac Newton was sitting) the rate of expansion began to slow.

The universe *was still expanding*, but not as fast as it had been at first. This slowing down continued for billions of years. (Until recently it was thought that the rate of expansion was still slowing down! It now appears, however, that this may not be the case. The reader may refer to pages 115-116 for more on this subject.)

So there is not, and never has been, any empty space out there waiting to be filled by an expanding universe. *The farthest limits of the universe also mark the farthest limits of space.*

Appendix (1), continued.

Because space is expanding this explains the so-called *red shift* of the light coming from bodies that are moving away from us.

When we examine light coming from these distant objects we find that all the various wavelengths are longer than they ought to be. The waves all appear to be *shifted* towards the *red* (the longest wave) end of the spectrum. We know this because of certain dark lines, gaps in the spectrum where some frequencies have been absorbed.

Individual elements absorb specific frequencies. So by studying the spectrum of light from a distant body we can tell which elements are present (see page 77, also the footnote on page 110). We can get an idea of the quantities too, as the broader and darker the line, the more of that particular element is present.

These dark lines do not shift, however, and it is by comparison to them that we can see the remaining waves have all become longer. Furthermore the *amount* of the red shift tells us the *speed* at which the source of the light is moving away from us.

If there were no dark lines in the spectrum then we'd have had no reference point against which to compare the rest. The waves might have shifted towards the red, *but there'd have been no way to tell.*

Let's leave light waves for a moment. You'll remember on page 19 we told how *sound* waves can be of different lengths, and that the shorter sound waves produce a higher note, and longer sound waves a lower note. And on page 21 we said how the length of the wave *received* will be *shorter* (producing a higher note) than the length *emitted* if the body emitting the sound is *approaching* the listener, and will be *longer* (producing a lower note) if the body emitting the sound is moving *away* from the listener. We said this was called the Doppler Effect, and that you'd find more about it in appendix (9).

Light waves are similar in this respect. When the body emitting the light is coming *towards* an observer the light waves received will be *shorter*, and more *violet* in color (p.22). And if the emitting body is moving *away* from the viewer then the waves will be *longer* and will have a *redder* color. The waves will be *shifted* towards the *red.*

So whether the body is moving *toward* an observer or *away* from him will determine whether the waves will be shorter or longer. And (especially) in the case of light (the speed of which is *absolute*) the *amount* by which the waves are shorter or longer will tell us the *speed* at which the emitting body is approaching or receding!

Appendix (2).

In chapter six, in explaining why the universe might not have a center, in terms that we can understand, we talked about a *fourth* dimension. By that we meant a fourth *spatial* dimension, that is to say a fourth dimension *in space*. This should not be confused with *time*, which is what most people first think of when they hear the dreaded words fourth dimension.

Now we've said that in this book we're not going to be getting involved much with general relativity, and the space/time continuum and things like that. But if you've ever tuned in to one of the handful of television programs aimed at viewers with intelligence quotients approaching triple digits, then you've probably heard people talking about *space/time*. So it might be appropriate here if we just examine for a moment why time should be considered by most people to be another dimension. Take the following example.

Suppose you are to meet with a very important person and are given exact instructions on how to get to the chosen meeting place. The meeting is to take place at noon on September 22nd. You set out clutching the piece of paper with the directions on how to get there.

Following the directions on the piece of paper you head south along the busy street until you reach the cross street mentioned in the directions. You don't need to go any farther north or south! *One of the dimensions, the north/south dimension, has been reached.*

A signpost indicates that the number you want is farther west, so you head west along the cross-street until you come to the building where the meeting is scheduled to take place. Now you need go no farther east or west. *Two of the dimensions have been met.*

According to the directions, you are to take the elevator to the fourth floor where the person you're to meet with will be waiting.

You step out of the elevator on the fourth floor. Now you have satisfied the *third*, the up-and-down, dimension. All three of the (*spatial*) dimensions have now been met! But even though you've followed the instructions there's no-one there to meet you!

You look at your watch. (It's the kind that tells you the date as well as the time.) The watch says it's noon all right, but on the 22nd of August not the 22nd of September. You're a month early! You had three of the dimensions right but not the *fourth* one!

In a very real sense then time is a *fourth dimension*. If you miss *any one* of the four dimensions then the meeting cannot take place!

Appendix (2), continued.

But the fourth dimension we were talking about in chapter six is not *time*. It's a *spatial* dimension, a fourth dimension *in space.*

A fourth dimension in space is something we can't visualize. It's an entirely unobservable dimension. Try and picture it in your mind, the way we picture the other dimensions. Most likely you can't. (Pages 43 & 44 are the best we could come up with.) But just because we can't visualize it doesn't mean its existence can't be deduced in other ways.

However it gets worse, for most scientists believe there are *more* than just four dimensions. One influential group, the string theorists, say there are a total of ten dimensions, one being *time* and the other nine being spatial. Another group, called the super gravity theorists, claim there's even more, and there's an *eleventh* dimension.

In recent developments it seems the latter view is starting to prevail, and that there are, in fact, eleven dimensions. Even the ten dimension people have now come around to thinking there might be an eleventh dimension after all, and claim there never really was any disagreement about it anyway!

All this talk about the existence of ten or eleven dimensions, like the notion that there may be an infinite number of parallel universes, seems to this author to be long on theory but short on evidence!

But in any case, if there are as many as ten or eleven dimensions, in this book when we ask you to think about a mere *four* dimensions just think how much worse it could have been!

Sometimes just when you think you've solved a problem new information comes to light that calls the solution into question. For while everyone agrees there has to be more than three dimensions, and differ only as to *how many*, an argument can be made that in fact there *can* only be *three* spatial dimensions or none of us would even be here! The argument goes as follows.

The sun radiates heat and light outwards. Now imagine that the sun is at the center of an enormous hollow sphere. And let's say the radius of the sphere is 93 million miles. Each square foot of the inside surface of the sphere would be receiving the same amount of energy currently being received by each square foot of the earth's surface (at least at a point where the sun was directly overhead) the earth also being 93 million miles away from the sun.

Appendix (2), continued.

Now suppose the radius of the sphere were to be *doubled.* Then, as we all know, the inside surface area of the hollow sphere would be *quadrupled!*

In that case the amount of energy received by each square foot of surface of the inside of the huge hollow sphere would be only *a quarter* of what it had been before. The radius of the sphere has *doubled*, but the amount of energy received by each square foot of the inside surface of the hollow sphere is reduced to *a quarter* not a half. (The propagation of heat, light, and *any* electromagnetic wave, follows an *inverse square* relationship!)*

Now, the argument goes, if there were *four* spatial dimensions, instead of three, then the surface area of a four dimensional sphere (whatever *that* would look like) would go up as the *cube* of its radius, not the *square*. Then, instead of an inverse *square*, we'd have an inverse *cube* relationship!

Double the radius and the amount of radiation received would go down by a factor of eight (2^3) not four (2^2). This would be bad news, because it would not permit stable orbits for the planets.

An inverse square law exactly balances centrifugal force and gravity. But in a four dimensional universe gravity would change more rapidly. (Scientists say it would have a "steeper gradient.")

If gravity had a steeper gradient then any tendency for a planet's orbit to stray would be instantly amplified. The planet would either plunge down into the sun, or spiral off into the depths of space.

So are there only three dimensions, or are there four or even more? It seems we haven't quite got it figured out yet!

As the garage mechanic said to the physician "there's still a lot we don't know about the 1998 Toyota."

* There is a *small* body of opinion that considers it *possible* that, in the case of gravity, it's not as simple as that, and that gravity might act differently at very great distances. This was mentioned briefly on page 116 (just over half way down).

Appendix (2), continued.

It should be mentioned that this idea that there can only be three dimensions is not one which receives much support from among the scientific community.

Meanwhile we hope you're finding all this interesting. Don't worry if you can't understand more than three dimensions, *nobody* can! But despite the problem described above, in the view of most scientists and other clever people there has to be more than three dimensions. Exercises like the one on pages 43 and 44 (as we just mentioned) show that while, like the flat people, we may not be able to picture an additional dimension in our minds, that doesn't stop us from concluding, based on a number of factors, that it must exist.

It's beyond our experience, that's all. The fourth dimension is a different *kind* of dimension, one we're not able to visualize. (And let's not even *think* about all the rest of the dimensions there's supposed to be!)

Appendix (3). Goldbach's Conjecture.

In 1742 a one time professor of mathematics and tutor to Czar Peter II of Russia, a man by the name of Christian Goldbach (1690-1764) noticed that even numbers could, or so it seemed, always be expressed as the sum of two *prime* numbers.

Let's try it ourselves and take the first even number we think of, say the number 38. That can be expressed as 7+31, or 19+19. Or there's 90, which can be expressed as the sum of 29+61, or 7+83, or 17+73 (and perhaps other ways too). All of these are prime numbers.

So you can see that there are in fact (almost) always *more than one* combination of prime numbers by which an even number can be expressed, the only known exceptions being the numbers 4, 6, 8 and 12 which are each the sum of only one such combination.

Why is this? Prime numbers make up only a small percentage of all numbers. They can only be divided by themselves or by one.

Our ancestors were superstitious about prime numbers. Today most tall buildings don't have a thirteenth floor. Oddly enough they always seem to have fifth, seventh and eleventh floors, which are also prime numbers but evidently not thought of as being particularly unlucky. Perhaps this is because thirteen comes right after twelve, and *that* can be divided by almost anything.

Prime numbers are less common among the higher numbers, but they still exist no matter how high you go.*

The only *even* prime number is 2. All the other prime numbers are *odd* numbers. Furthermore 2 is also the only even number for which Goldbach's conjecture doesn't work. However, this apparent exception is only because, for some odd reason, one not considered to be a prime number. So it doesn't really invalidate the conjecture.

Unable to find an even number (other than two) that could *not* be expressed as the sum of two prime numbers Goldbach handed the problem to Leonhard Euyler (1707-1783) the greatest mathematician of his day. (Engineers know Euyler for his method of figuring the compressive strength of struts.) Euyler said he'd look into it.

* The biggest prime number found so far is ($2^{859433}-1$). This huge number has no less than 258,716 digits! If you were to print out this number, filling page after page with all those digits, you'd finish up with a book almost as big as the one you're reading! And yet it can't be divided by *anything!* (Except by one or by itself, of course.)

Appendix (3), continued.

What was needed was a *proof*, a logical explanation of *why* this should be so. Alternatively, all that was needed was just *one single exception* to the rule! This would mean that the others were all just coincidences.

Euyler, and just about every other mathematician for the last couple of hundred years or so, have tried to either *prove* or *disprove* this apparent rule. Meanwhile we have to keep calling it a conjecture until someone can either come up with a satisfactory proof, or point to a single case where it doesn't work.

There've been no developments since the time of Goldbach and Euyler. More recently, by using computers (you remember, when computers actually used to *compute*) mathematicians have scanned billions and billions* of even numbers and have yet to find a single even number that *cannot* be expressed as the sum of two prime numbers. Nor has anyone ever been able to come up with a *proof* of *why* this should be so.

And there the matter stands today, a continuing challenge to all mathematicians. But here it seems the human brain has come up against a problem beyond its ability to resolve.

It's likely the mystery will never be solved now. No-one much tries anymore.

I give up!

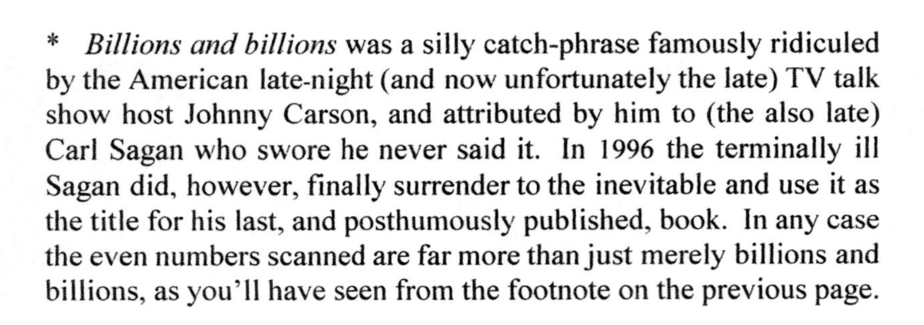

* *Billions and billions* was a silly catch-phrase famously ridiculed by the American late-night (and now unfortunately the late) TV talk show host Johnny Carson, and attributed by him to (the also late) Carl Sagan who swore he never said it. In 1996 the terminally ill Sagan did, however, finally surrender to the inevitable and use it as the title for his last, and posthumously published, book. In any case the even numbers scanned are far more than just merely billions and billions, as you'll have seen from the footnote on the previous page.

Appendix (4).

On pages 65 & 66 we discussed the case of an astronaut who traveled out into space then back again. On his return he found that, despite their both having been born on the same day, the twin brother he'd left behind on earth was now ten years older than he was!

Now according to Galileo, if the astronaut was moving *relative to his twin brother*, it would have been just as true to say the twin brother was moving *relative to the astronaut!* Why then, was it the twin brother who aged faster? Why couldn't it just as well have been the other way round? This is the "twin paradox."

Well suppose the astronaut, instead of returning to earth, had continued in a straight line. Then the two brothers' journeys would have been *symmetrical.* After any long enough period of time either one of them could equally well have claimed it was the other who was younger. But there would have been no paradox, for the twins could never have met up again to compare notes.*

However, the astronaut did *not* travel in a straight line. After a while he had to turn around and head back while his twin brother, back here on earth, continued to move through space/time at a steady speed and in a straight line!

Now let's pause here for a moment, because you may think if time were passing *more slowly* in the space capsule (as seen from back on earth) while the astronaut was *going away*, then when he was *coming back* the situation would be reversed and for the return part of the journey time in the space capsule *would be passing more quickly*. If this were the case then the time distortions would have cancelled each other out, and when the two brothers met up again they'd both be the same age.

However, this is not the case. Whether the astronaut is going away or coming back *doesn't make any difference.* This will be better understood if the reader refers to the description of Feynman's clock in chapter eight. You will see it's only the *relative speed* that matters. The *direction* of travel makes no difference at all!

* In a way they'd have *both* been right! By the time the two brothers were far enough apart, and because it would take light *time* to travel this distance from the one brother's face to the other brother's eye, by the time the image reached *either one of them* (and supposing they could see that far) for every light year (distance) the brothers were apart, each would see the other as being one year younger.

Appendix (4), continued.

Now we've mentioned several times that for special relativity to apply things have to be traveling *at a steady speed and in a straight line.* Special relativity collapses if either of the bodies involved is *accelerating!* In fact the reason Einstein called it the *special* theory was because it deals *only* with the *special* case where bodies are moving at constant velocity and in a straight line. Special relativity *does not apply* when dealing with phenomena observed by a person who is *accelerating* relative to the observed system, or vice-versa.

So the first thing we have to do is find out if either the astronaut or his twin brother was accelerating.

On page 50 we said that, simply stated, acceleration is the rate at which the speed of something is changing. (We usually mean going faster.) But in reality it is more than that. To engineers acceleration is a change in speed *in either magnitude or direction!*

The astronaut traveled out into *space then turned round and came back.* Of course his direction couldn't be reversed instantly like that. He would have had to slow down, come to a stop, then accelerate away in the opposite direction. (Notice we use the word *accelerate!*)

But even if the astronaut *could* reverse course instantly, or if we ignored the slowing down and speeding up parts of the journey, or even if he didn't slow down at all and just swung around in a wide U-turn, he would still be accelerating for the simple reason that *he changed direction!*

A race car driver going round a bend, even if he maintains a constant speed, because he's going round a bend and *his direction is changing* then he is *accelerating.*

To maintain a steady speed in a straight line doesn't require a force to be applied (page 83). *Acceleration,* on the other hand, *does* require a force ($F = ma$, remember).

If you were to swing a rock (at a steady speed) on the end of a piece of rope, the force needed to *accelerate* the rock (in this case to change its direction) is being transmitted to the rock through the tension in the rope. If the rock were to come free from the rope it would continue to move, *but now at a steady speed and in a straight line* which (as we just said) doesn't *require* a force to be applied!

Galileo said you only need to apply a force *if you want to change a body's motion in some way* (in either magnitude *or* direction).

Appendix (4), continued.

So we've established that the twin who stayed behind on earth traveled through space/time at a steady speed and in a straight line while the *astronaut's* journey involved accelerating! Because of the fact that one of the bodies was *accelerating*, then special relativity no longer applies. Einstein's first postulate has not been satisfied.

So now we need to think in terms of space/time, and in *general* relativity there's another law (one we're not going to get into any more than is necessary) which says that time, in a frame of reference that is *accelerating*, will pass more slowly when observed from a frame of reference that is *not* accelerating. (See below.)

To better understand the situation it might help to think of the twin who remained behind on earth (and who was moving through space/time at a steady speed and in a straight line) as having traveled along one side of a triangle.

But the astronaut, although both brothers eventually ended up at the same place, *got there by a longer route*. Because he traveled out into space then turned around and headed back to rejoin his brother who, meanwhile, as we said, had been traveling through space/time (together with the earth) at a steady speed and in a straight line.

The astronaut had to travel a greater distance. Therefore we're going to say the astronaut traveled along *the other two sides* of *that same triangle*. (He would, it's true, have covered more miles.)

However, in terms of space/time, the extra *distance* is *exactly balanced* by the fact that *time,* in the astronaut's frame of reference, *because he was accelerating*, was seen to be passing more slowly. This, as we explained above, is a feature of *general* relativity which we're not going to get into.

As a result each brother, from start to finish of their respective journeys, will have "used up" the same amount of space/time.

Has this not been the case they'd not have been able to meet up again. As it was, at the end of the trip both brothers found themselves *occupying the same position in the "space/time continuum"*.

But if we concern ourselves *only* with the *time* component, time as measured by clocks (or by the body's aging process) instead of *space*/time, then there's nothing to offset the slowing down of time in the space capsule (as seen by his brother back on earth). *Whatever distances the brothers may or may not have traveled to get to where they are is no longer either measurable nor relevant!*

Appendix (4), continued.

All that matters now is that, due to the slowing down of time in the space capsule, caused by its *acceleration*, as viewed from back on earth, when the brothers meet up again the one who stayed at home will see the returning astronaut as being younger. (As we said, it doesn't work the other way around.) The twins, despite their both having been born on the same day, are no longer the same age!

Now we've shown (we hope) how *time* can pass more slowly in one frame of reference than another. Theoretically, if a body could reach the speed of light (relative to another frame of reference) then time, seen from that other frame of reference, *would stop altogether*.

But can time ever go *backwards?* (You don't need to read the rest of this if you don't want. It's of general interest only, and not that important.)

Television programs such as Star Trek will sometimes feature episodes where, to make a good story, characters find themselves transported back in time. If enough programs of this type are shown then a lot of people will come to believe such things are possible. After all, half the population already believe we're being visited by flying saucers carrying aliens from another world* (who take the greatest care not to leave behind any tangible evidence of their visits).

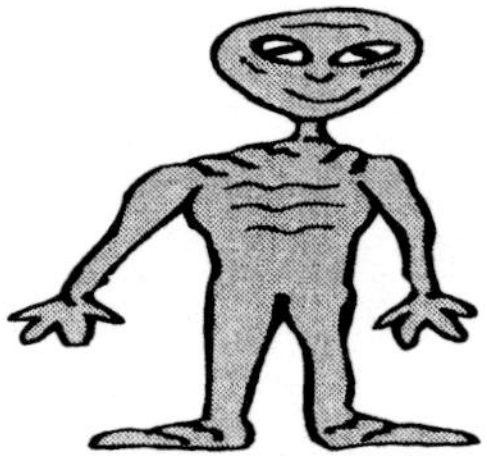

Typical alien from
another planet.
(Usually colored green.)

"*The moving finger writes, and having writ moves on,*" wrote Omar Khayyám, the twelfth century tippling tentmaker of Teheran (beneath the bough, and with a jug of wine, no doubt). He evidently didn't think time could go backwards.

Most scientists think that, while the rate at which time flows may vary, it flows in only one direction. The second law of thermodynamics is sometimes quoted as an example of this. This law says that heat always flows from a hotter body to a cooler one.

* Which is not surprising, when you consider there are still a lot of people out there who don't believe in evolution, and think we were all created somehow looking much the way we look today.

Appendix (4), continued.

If you introduce an ice cube into a cup of hot coffee you will observe the ice cube melt and the coffee become cooler. But you will never observe a cup of cool coffee become hotter by drawing heat from an ice cube simultaneously forming within it, even though the total *mass/energy* of the ice and the coffee would be unaffected, so there's really no reason why this shouldn't be possible.

The reason heat won't flow from a cold body to a hotter one is because, though we may not know why, something is always present in nature when a process is spontaneous. That something is entropy.

Entropy introduces an irreversibility, an *asymmetry* into the equation. *In a closed system entropy must either remain unchanged or increase.* It is a measure of the *disorder* of the system, and if order is to be restored additional work is required. Work which nature, in the normal course of events, will not provide.

Consider for example a new deck of playing cards which, when first opened, are arranged exactly in numerical order and by suit. If thoroughly shuffled the cards will become all mixed up and in no particular sequence. Similarly a china vase falling from a shelf will break into pieces. To restore these things to their original state requires an effort that nature, as we said, will not provide.

But entropy is not the only reason why a spontaneous process is, for all intents and purposes, irreversible. We also have to consider the *probability* of an event occurring.

Consider again the deck of cards we just mentioned above.

We said the cards were originally arranged in exact numerical order and by suit. If the cards were to be shuffled, we said, they'd then be all mixed up and in no particular sequence.

In fact, once the cards have been properly shuffled, for a dealer in a casino (or anywhere else) to deal them out in their original order the probability is less than one chance in 10^{68}, or ten to the sixty-eighth power! This is a ten with 68 zeros after it, a figure beyond our ability to comprehend. (The author figured it to be about $8{\cdot}22 \times 10^{68}$.)

Appendix (4), continued.

If the odds are only one chance in 10^{68} * that means if every man, woman and child *who has ever been born*** were each of them to take a deck of cards and start shuffling them, without ever stopping or slowing down for any reason then (on average) the length of time it would take before any of the decks of cards ever again appeared in their original order would be trillions of times longer than the universe has even existed! (And because this is only an average it could take even longer!)

Although entropy can sometimes be reversed by intelligent life we never observe, in nature unaided, other than by the acts of such intelligent life, progression towards a more complex condition. At least not here on earth. We all know this, at least subconsciously. If one is shown a motion picture of bits of broken china assembling themselves to form an undamaged vase, it will immediately be recognized that the movie is being run *backwards.*

* Sometimes in studying the cosmos we come across even larger numbers. For example there's the number 10^{76}, or ten multiplied by itself seventy-five times. This is an estimate of the total number of *atoms* in the universe, given in a recent lecture by the director of the Vatican Observatory. (And *his* sources ought to be good!)

Another authority says the number of elementary particles in the universe is about 10^{80}. (We take this to mean protons, neutrons and electrons.) As 99·9 % of the universe is made up of either hydrogen or helium, each of which consist of only a few elementary particles (two and six respectively) the two figures are not really compatible!

** Scientists estimate this to be about seventy billion people, over the approximately 200,000 years since the human race reached more or less our present stage of evolutionary development.

Appendix (5).

We've shown how mass can be changed into huge amounts of energy. But when an atomic nucleus is split we're only able to turn the tiniest part of the mass into energy.

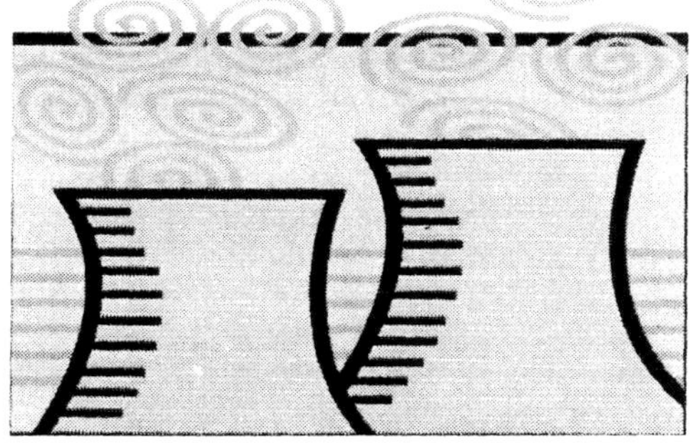

What if we could turn *all* the mass into energy? Every bit of it! Then how much energy would we get?

Let's consider a pound of something that we're going to change into energy. All of it, the whole pound. (And remember, it could be a pound of *anything!*)

Because the speed of light is so great, and since the *energy* we get is going to be proportional to the *square* of this speed, then the amount of energy released will be unimaginably great.

We all know how a nuclear bomb can liberate vast amounts of energy. Even a small one destroyed the city of Hiroshima and killed a hundred thousand Japanese, mostly innocent civilians, with tens of thousands more dying later in agony from burns or radiation. *Yet only a tiny fraction of the uranium mass was changed into energy.*

On a recent television program* the narrator stated that in the Hiroshima bomb it was only *six-tenths of a gram* of mass that was converted into energy. (By this author's calculations it must have been more than this.) But either way it was only the *tiniest fraction* of the original fissile material that was changed into energy. The rest of it finished up as atoms of lighter elements, not energy.

* You can't believe everything you see on television. On the other hand there's little reliable information available to the general public on such things as the yield of a bomb, how much fissionable material it takes to make a critical mass, what is the heating value of that explosive by which the yield of a bomb is conventionally expressed, and so on. It's not a subject those in the know like to talk about.

Appendix (5), continued.

That only a tiny part of the mass is changed into energy is also true of nuclear fusion. (You'll find reference to this on pages 78-79.)

But be that as it may, we were debating how much energy we'd get if we *were* able to turn *all* of the mass into energy.

If a whole pound of mass could be turned into energy it would yield about *thirty-thousand-trillion* foot pounds of energy. A trillion is a 1 with twelve zeros after it, or a million times a million. We can write this as 10^{12}. All this from just one pound of mass, remember!

This is equal to about 38½ x 10^{12} BTUs of heat. If you had a one-kilowatt electric heater in your living room, the *energy* in this pound of mass would be enough to keep the heater burning for the next one million two hundred and ninety thousand years. To put that in perspective, it was about that length of time ago that our ancestors were first starting to come down from the trees!

The energy in a pound of mass is equal to the output of a major power station in more than a year. All this from only one pound of mass! And the fuel wouldn't have to be oil, it could be *anything!*

The energy that you'd get from the *complete* conversion of the *mass* of a pound of oil into *energy* would be more than *two billion* (two thousand-million) times the energy we get from burning it the way we do now.*

In a few years, when we run out of oil** and the lights go out, it won't be so fashionable to oppose nuclear power. Unfortunately by then it will be too late to save the planet!

* Of course the *complete* conversion of mass into energy is not something we're likely to achieve any time soon! *This* doesn't even happen inside the sun.

** While this book was being written the price of gasoline at the pump roughly doubled.

Appendix (6).

On page 89 we said Einstein's value for the *total* energy of a body was given by the equation:

$$E = \frac{mc^2}{\sqrt{1 - v^2/c^2}} \quad *$$

Then by the time we'd arrived at page 93 we'd shown how we thought Einstein came up with this.

Maybe we're missing something, because it seems if Einstein was right then we already had everything we needed and could have finished this book right there. There'd have been no need for chapter fourteen even.

Because if we wanted to see how much *energy* was contained in the *mass* of a *stationary* body (a body that's sitting on the table and not going anywhere) then surely this was a simple matter. All we had to do was take Einstein's equation and say that the body's velocity was zero.

Because if v is *zero* then Einstein's equation gives us:

$$E = \frac{mc^2}{\sqrt{1 - v^2/c^2}} = \frac{mc^2}{\sqrt{1 - 0^2/c^2}} = \frac{mc^2}{\sqrt{1 - 0/c^2}} = \frac{mc^2}{\sqrt{1-0}}, = \frac{mc^2}{1}, \text{ which is } mc^2.$$

So $E = mc^2$. Simple as that! Yet Einstein himself in his 1920 book went into the lengthy and seemingly unnecessary discussion, converting his equation into a series, then analyzing each step of the series and so on.

Perhaps this is what Einstein meant when he told his secretary years later that he could have put it more simply. Or maybe there's more to it than that. As we said in the foreword to this book, your author never claimed to be an expert (which is why you've been able to follow everything so far).

* Don't forget that E is the *total* energy of the body, the energy it has that is represented by its *mass* when it was not moving (relative to the observer) *plus* whatever additional mass/energy it possesses due to any motion it might have.

Appendix (7).

One of the times we come across *sines* is when we're dealing with *right angle triangles.*

Right angle triangles are those where *one* of the angles is ninety degrees (90°). Below is a right angle triangle the lengths of whose sides in this case happen to be the whole numbers 3, 4 & 5.

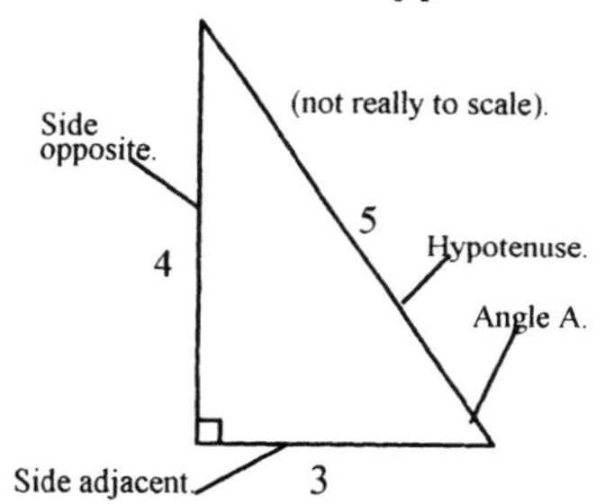

This particuler triagle happens to be a three-four-five triangle, so named because of the lengths of its sides.

Because it is a three-four-five triangle angle A will be about 53° and the other one (the one at the top) about 37°.

There are plenty of other right angle triangles whose sides are whole numbers. There's 5, 12 &13 for example ($5^2 + 12^2 = 13^2$).

There's lots of others too. The Babylonians made a list of some of them thousands of years before Christ.* Today, using computers, we have found a lot more.

You'll see that the right angle is at the bottom left. The side opposite is called the hypotenuse. (You'll know that from school.)

If we take angle A then we'll see that the *other two sides* of the triangle are the side *opposite* and the side *adjacent to* angle A.

An angle has certain properties. One of these is its *sine*. To calculate the sine divide the *side opposite* by the *hypotenuse*. In this case divide 4 by 5, which is 0·8. The *sine* of angle A is 0·8.

In the above triangle the value of angle A is sine^{-1} 0·8. (Sine^{-1} is the way engineers say "the angle whose sine is.")

It works out that A is just over 53°. Because the angles of a triangle must always add up to 180° this means that the other angle (the one at the top) must be just under 37°.

On the next page we show another interesting triangle.

* Or before the common era, as we're supposed to say these days!

Appendix (7), continued.

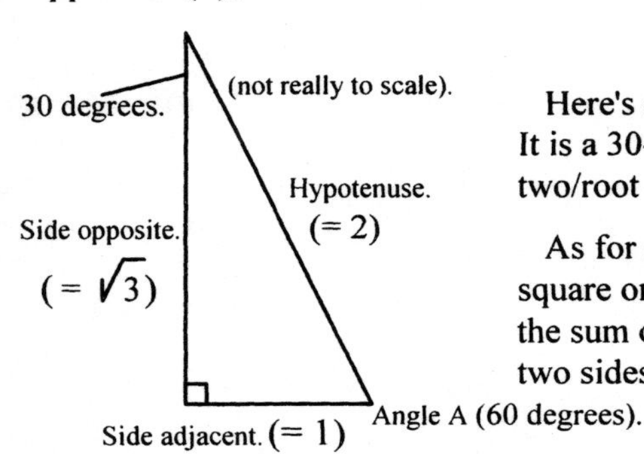

Here's another interesting triangle. It is a 30-60-90 triangle, or a "one/two/root three" triangle.

As for all right-angle triangles the square on the hypotenuse is equal to the sum of the squares on the other two sides.

Here the sine of angle A is the square root of three, divided by two, which is 0·866.

Apart from its *sine* an angle also has other properties. Some of these are its *tan* (short for tangent), its *cos* (or cosine), but we don't need to concern ourselves with any of these right now.

You will see from the sketch that if angle A were almost 90° then the base of the triangle would be very small and the side opposite to angle A would be very large, until it was *practically the same length as the hypotenuse*.

So that if A were actually 90° its *sine* would be one. (Sines can never be *greater* than one.)

And if angle A were *very small*, then the side opposite it would be very small too, compared to the hypotenuse and the adjacent side, and therefore its *sine* would be practically *zero*.

Here are some values for the sines of various angles.

Angle	0	30	45	60	90	120	135	150	180
Sine	0	0.5	0.7071	0.866	1.0	0.866	0.7071	0.5	0

If you plot these values on a graph you'll have a *sine wave* like the one we showed you on page 30.

In designing generators engineers try to produce as near to a perfect *sine wave* as possible. This is important in balancing a three-phase electrical system.

Appendix (8).

On page 69 we said that to accelerate a space ship to *half* the speed of light would require a rocket so big that, when it blasted off, it would incinerate everything within hundreds of miles. This was because of the huge amount of (kinetic) energy a body would possess when traveling at half the speed of light. And this energy, of course, would have to have been provided by the rocket.

To find out how much energy would be needed to accelerate the capsule to only *half* the speed of light we'll make use of Einstein's equation for the total energy, $E = \dfrac{mc^2}{\sqrt{1 - \frac{v^2}{c^2}}}$.

When Apollo Eleven left the launch pad it weighed nearly four thousand tons. But most of this was in the first and second stage booster rockets which soon fell away. In our case it would be a relatively small mass, just the command module, that reaches design speed. How much would this module be likely to weigh?

To reach a planet of another star at half the speed of light would take eight and a half years to get to even the *nearest* star system. And another eight and a half years to get back. (Only how would we *do* that?) That would be seventeen years in just travel time alone!

Due to the length of time the astronauts would have to be away, in order to carry all the necessary personnel, equipment, food, fuel etc. the space ship would need to be quite big, probably weighing around 100 tons *at least!* (Probably much more!)

Now if the *mass* of the space capsule was 100 tons then its total energy at half the speed of light would be $E = \dfrac{mc^2}{\sqrt{1 - \frac{v^2}{c^2}}}$.

$$\text{Or } E = \frac{\left\{100 \text{ tons} \times 2{,}000 \dfrac{\text{lb.}}{\text{ton}}\right\} \times \left\{186{,}000 \dfrac{\text{mile}}{\text{sec.}} \times 5{,}280 \dfrac{\text{ft.}}{\text{mile}}\right\}^2 \times \left\{\dfrac{\text{lb. sec}^2.}{\text{lb. } 32 \cdot 2 \text{ ft.}}\right\}^*}{\sqrt{1 - \left\{\dfrac{(\frac{1}{2} c)^2}{c^2}\right\}}}$$

This works out to $6{\cdot}94 \times 10^{21}$ foot pounds and is the *total energy* of the capsule.

* The last item is a correction factor, see middle of page 151.

Appendix (8), continued.

However, the capsule already *had* a mass of 100 tons before it blasted off. Using $E=mc^2$ we can calculate that this stationary mass, would be $6{\cdot}01 \times 10^{21}$ ft.lbs. So that the *additional* energy needed to accelerate the space capsule to half the speed of light would be:

$(6{\cdot}94 \times 10^{21})$ minus $(6{\cdot}01 \times 10^{21})$ or $0{\cdot}93 \times 10^{21}$ ft.lbs.

We can write this as 930 *million-million-million* foot pounds. In terms of heat this equals 1·2 *million-million-million* BTUs.

If we used a rocket to launch the space vehicle (which at the present time is the only way we know how to launch space vehicles) then it would take 32 *billion* tons of rocket fuel (depending on the kind of fuel used).

The figures involved are almost beyond our imagination! It's more like what takes place inside a star! (It's *astronomical!*)

What this means is that travel to planets of other stars is, for all intents and purposes, *impossible!* Exotic things like worm holes through space are nothing more than wishful thinking at this point. They enable the star ship Enterprise to visit distant galaxies during a half hour television program, but that's about all! (You can tell this author is no great believer in flying saucers.)

Of course things that seem impossible are sometimes found later to be possible after all. Who knows, there might be some big new breakthrough just around the corner. There have, for example, been investigations into the feasibility of harnessing nuclear power for the launching of space vehicles. Something of this nature would appear to be necessary if we are ever to meet the rocket's enormous energy demands.

It has also been proposed that a spaceship could be propelled by a huge sail, possibly as much as several square kilometers in area and driven, not by wind, but by *light* from the sun. The tiny *photons* of light would accelerate the spaceship forward with miniscule force, one *quantum* of energy coming from each photon (even though they have no mass).

A problem here is that, due to the smallness of the energy being imparted by the impact of the tiny photons, the rate of acceleration would be extremely low. By the time the speed of the spacecraft had reached a useful level the sun would have been left so far behind that there'd no longer be enough light to have much effect.

Appendix (8), continued.

Another method being tried is ion drive. A mission is currently under way utilizing this method of propulsion for the first time. The acceleration produced by the ion drive, like that of the photon sail, will be so small that even after several days it will have reached a speed of only about sixty miles an hour. However the force, like the force produced by the photons, will be continuous, so that eventually an enormous speed could be attained. And with ion drive the ever increasing distance from the sun would not be a limiting factor. But we don't want to get too far away from the subject here!

Still, absent some revolutionary new concept, judging by past experience, and with even the most optimistic of forecasts for future advances in space technology, it would seem that, apart from the occasional trip to the moon, and probably Mars in due course, the human race will be confined to, and eventually die with, planet earth (or almost certainly billions of years sooner).

So far we've not found any other body in the universe capable of supporting life as we know it, nor even one that might be adapted to such purpose. Some of the things we see on television on the subject are absurd.

As to looking further a-field, well let's just say the outlook is not promising. Once you begin to comprehend the mind boggling distances involved you'll realize that travel within the solar system is one thing, but interstellar travel is a whole different story!

And finally, we said that when calculating the value of mc^2 it was necessary to use the conversion factor $\left\{\frac{\text{lb.sec}^2}{\text{lb.32·2 ft.}}\right\}$.

This is because the units for *weight* and *mass,* adopted centuries ago and still in use today, were not well chosen.

The trouble is that mass and weight are not the same thing. Back then a body's *mass* was considered to be simply what it weighed! If it weighed more then it had more mass they said (which is correct as far as it goes). However, they failed to consider that measuring the same body's *weight* on another celestial body, where gravity was not the same as on earth, would have given a different reading.

The *weight* of a body is determined by the strength of the gravitational *force* to which it is subjected, while the body's *mass* is constant (well near enough, unless you're accelerating it to very high speed that is) and quite unaffected by gravity, or by anything else.

Appendix (8), continued.

And since we've mentioned bodies being *accelerated* to very high speeds, it should be understood that the increase in the (*inertial*) mass of a body due to this acceleration is only significant at speeds in the *thousands of miles per second* range!

When we send a spaceship to the moon it's moving many times faster than the fastest rifle bullet so its (inertial) mass increases. Yet this is such a (relatively) slow speed that the increase in mass is too small almost to calculate (at least when you've been out of school for as long as this author). But the increase in mass is certainly far less than a millionth of the original mass! You'd have to be moving thousands of times faster before the increase in mass was big enough to make any difference!

We were talking about *weight*, weren't we. Well in Newton's time people weren't giving a whole lot of thought to what things might weigh on other celestial bodies. They were very sensibly concerning themselves with more interesting matters, like apples falling from trees. (Incidentally Newton never said an apple hit him on the head. His ideas on gravity, he wrote later, came to him "*while I was in a contemplative mood, and was occasioned by the falling of an apple.*" Nothing about it hitting him on the head!)

But as far as most people were concerned *mass* and *weight* were much the same thing, so the same units could be used for both. As a result unit *force*, in pounds, acting on unit *mass*, in a different kind of pounds, does *not* cause unit acceleration.

For example Ohms law (V= IR) says that a potential difference (or voltage) of <u>one</u> volt will cause a current of <u>one</u> amp to flow in a circuit of resistance <u>one</u> ohm. Another law says that if you give <u>one</u> calorie of heat to <u>one</u> cubic centimeter of water this will raise the temperature of the water by <u>one</u> degree Celsius. Similarly <u>one</u> BTU of heat will raise the temperature of <u>one</u> pound of water by <u>one</u> degree Fahrenheit, and so on. This is how most laws work! But due to our ancestors' failure to take into account the fundamental difference between *mass* and *weight* this orderly state of affairs is unfortunately *not* the case for Newton's law of F=ma.

Appendix (8), continued.

If a *force* of one pound acts on a *mass* of one pound it *does not* cause an acceleration of one ft/sec². Unhappily for engineering students it causes an acceleration of 32·2 ft/sec².*

Being stuck with our current system of measurements, where we say that a body's *mass* is just what it happens to *weigh*, now the only way things would ever work out nicely is if we were living on a little planet where gravity was 32·2 times weaker than on earth, where things would *weigh* a lot less. Then if we tried to *accelerate* something by applying a *force* equal to what it *weighed*, then force really *would* equal mass times acceleration.

To put it another way, the value of g would then be only one foot per second per second, or $1\ {}^{ft}/_{sec^2}$ instead of $32 \cdot 2\ {}^{ft}/_{sec^2}$.

However if nothing weighed very much then (with the possible exception of the airline industry) this might create more problems than it solves.

Over the years engineers have devised many ways to correct this unfortunate state of affairs. When this author left the British navy after World War Two and entered college the correction factor used then was the one in the equation at the bottom of page 149, which was $\left\{\frac{\text{lb.sec}^2}{\text{lb.32·2 ft.}}\right\}$. (See also page 151.)

Earlier, while the author was still in the navy, a slightly different method was used. When you came across a *pound* you had to decide whether it was a pound *mass*, which was called a *little* pound, or a pound *weight*, which was called a *big* pound. A little pound (also sometimes known as a "poundal") was equal to a pound *weight* divided by "g" the acceleration due to gravity.

$$\text{Or } m = \frac{W}{g}.$$

For example, instead of saying F = ma you'd say $F = (\frac{W}{g})\,a$.

Or if you wanted to calculate the kinetic energy of a body then instead of saying it was equal to $\frac{1}{2}mv^2$ you said it was $\frac{1}{2}(\frac{W}{g})\,v^2$.

* On page 86 we said this is sometimes called Newton's Constant.

Appendix (8), continued.

If we'd used this method back on page 149, then instead of saying $E = \dfrac{mc^2}{\sqrt{1 - v^2/c^2}}$ we'd have said $E = \dfrac{Wc^2}{g \times \sqrt{1 - v^2/c^2}}$.

That way we wouldn't have needed to use a conversion factor because it would have already been built in.

What this says is that one pound of *mass* (m) = $\dfrac{1 \text{ lb. } \textit{weight} \text{ (W)}}{32 \cdot 2 \text{ ft}/\text{sec}^2}$.

And if we tidy this up a bit we get $1 \text{ lb.(m)} = \dfrac{1 \text{ lb.(W)} \times \text{sec}^2}{32 \cdot 2 \text{ ft.}}$.

This is where we get the conversion factor, because now, if we divide each side by *pound mass* we get $1 = \dfrac{\text{lb. sec}^2}{\text{lb. } 32 \cdot 2 \text{ ft.}}$, where the pound in the *numerator* of this new equation is pound *weight,* and the pound in the *denominator* is pound *mass*. (Of course both these methods are essentially doing the same thing.)

You'll especially note that on page 149, where this conversion factor is used, and where the energy is measured in *foot pounds,* if the conversion factor had *not* been used the *units* would not have come out right. In place of foot pounds we'd have finished up with units of $\dfrac{\text{lb.ft}^2}{\text{sec}^2}$ (which wouldn't have made any kind of sense).

No doubt by now the reader is thoroughly confused. Still, unless you plan to study engineering, which is not recommended (engineers don't enjoy much in the way of status or financial reward) then you probably don't need to worry about it. The distinction between mass and weight is examined here merely to accommodate the unlikely event that anyone may actually be interested.

Appendix (9). The Doppler Effect.

Imagine you're standing near a railroad track and a locomotive is coming towards you. The engineer is blowing the whistle which is emitting sound waves in all directions. Let's just worry about the waves that are coming in *your* direction.

When you hear a sound the *pitch,* whether it's a high note or a low note, will depend on how many waves are reaching your ear each second. As the *speed* of the waves is fixed (for any given set of conditions) the *number* of waves reaching your ear each second (the *frequency*) will depend on the *distance* (the *wave length*) from one wave to the next.

If the train's not moving there'll be a certain *distance* between the waves. If the whistle's pitch was the same as that of the middle C note on a piano then the distance between the waves would be four feet. (Or near enough for our purposes.)

But if the train were moving towards you then the *distance* between waves reaching your ear would be *less* than four feet. The *wave length* would be less, which we'll now explain.

The waves are traveling *through the air* at 760 miles per hour. Let's say a wave has just left the whistle and set out on its journey to your ear. Because in this case the pitch of the whistle is middle C, which means it has a *frequency* of about 263 waves per second, then the next wave will follow one two-hundred-and-sixty-third of a second later. Because the first wave is moving at 760 miles per hour the next wave, the one following it, will be about four feet behind the first one.

But in the interval between the first wave and the next wave *the train* has *also* been moving. The train has been chasing after the wave that it's just emitted, so that it's not as far behind the first wave as would have been the case *had the train not been moving*. Because of this the distance from the first wave to the next one, in other words the *wave length,* is actually going to be a bit *less* than four feet. Now that makes sense doesn't it?

Appendix (9), continued.

So the waves reaching your ear are now less than four feet apart. They'll still be traveling at the same speed but now they are closer together so that *more of them will be arriving.* Their *frequency* will be higher. Instead of 263 waves arriving at your ear every second you'll now be receiving perhaps 270 or 280. To *you* the whistle will have a *higher* note than middle C.

You might say that as each wave is created the train chases after it while emitting the next one, so that the waves get bunched up, or squeezed together, in front of the train.

This bunching up, or squeezing together, of the sound waves is the reason for what we all know as the *sound barrier.* If the thing that's emitting the sound is going *almost as fast as the sound waves themselves* then the waves get so bunched up in front of it that the airplane (or whatever it is that's emitting the sound) can be stressed to the point of breaking up (which happened sometimes).

The author remembers the death of Geoffrey de Havilland in an experimental British jet, the DH-108, manufactured by his father's company, attempting to break the sound barrier. Near supersonic speed the plane broke into pieces. Later a movie was based on it, *The Sound Barrier*, with Ralph Richardson as head of the company (though they didn't actually call him de Havilland) and Nigel Patrick as the pilot who died.

The sound waves being bunched up together, when they all reach the ground at the same time it sounds like an explosion. This is the *sonic boom* (which is actually two booms following in quick succession) and it's the reason why supersonic passenger aircraft like the Concord are not allowed to fly over densely populated areas. At least not at supersonic speeds.

Now as the train passes you and is moving *away* the *distance* between the waves will be *stretched out* instead of being bunched up, and the waves, instead of two-hundred and sixty-three of them per second, will now be arriving at your ear slightly more *slowly,* perhaps now only around 240 waves per second. Their *frequency* will be lower. You'll hear a note *lower* than middle C.

Most of us have experienced this Doppler Effect at one time or another, perhaps at a race track, like the Indianapolis 500, say, when a race car whizzes past. Or if you're trying to fix a flat tire by the side of the freeway while the traffic is speeding by.

Appendix (9), continued.

Remember, it's not the *speed* of the waves that changes when the locomotive is approaching the listener but the *distance between the waves*. And because the waves are traveling at the same *speed,* and the *distance* between them *is less,* then their *frequency* will be higher. (And the opposite when the train is *going away* from you.)

Another way to look at it would be to say that, because the train is *moving towards you* as it whistles, then each sound wave, in order to reach your ear now has a slightly *shorter* distance to travel than did the wave that went before it. Wave *two* will have a slightly shorter distance to travel than wave *one*. And wave *three* will have a slightly shorter distance to travel than wave *two.* And so on.

As it has a shorter distance to travel each wave will take a slightly shorter *time* to reach your ear than did the one that went before it. So while the waves are being *produced* at the rate of 263 waves per second (if we're talking about the middle C note, that is) *thanks to the movement of the train* each wave is now going to be slightly closer to the one in front of it *than it would otherwise have been.* And because they're all moving at the same speed if the *wave length* is *shorter* the waves will be arriving at a faster rate. There'll be *more waves* arriving per second, so the *frequency* will be *higher* and you'll hear a higher note.

As a matter of interest, just *what is* the distance between the waves going to be?

Well if the train (and its whistle) were *not* moving (relative to the person doing the listening) then the wave length (the distance between the waves as observed by the listener) would be the *speed* of the wave multiplied by the *time* between the emission of one wave to the emission of the next. That is to say *the distance traveled* at 760 miles per hour *in one two-hundred-and-sixty-third of a second.* As we said, this is about 4 feet. (Actually just under four feet three inches but we're calling it four feet.)

Appendix (9), continued.

But *if the train is moving towards the listener* then the distance between the waves, as they arrive at his ear, will be...

(a)...the distance traveled by the wave at 760 miles per hour in one two-hundred-and-sixty-third of a second (or 4 ft).

less

(b)...the distance traveled *by the train* during that same period.

Meaning the wave length now is *less than* 4 ft and you'll hear a higher note.

And if the train is *going away from you,* then the length of each wave, instead of (a) *minus* (b) will be (a) *plus* (b). This will cause the listener to hear a *lower* note.

If this explanation is confusing then perhaps the diagrams on the next two pages may help. They are sketches of the train, which is blowing its whistle as it approaches the man standing beside the railroad track.

Once again to keep things simple we're going to say the pitch of the train whistle is the note middle C, a note which has a frequency of 263 waves per second. The train at this point is not moving. The *distance* between the man and the train is 64 feet.

At the speed of sound it will take *one sixteenth* of a second for the *first* wave to travel the 64 feet from the whistle to the man's ear. At this frequency *if the train is not moving* then the waves will be 4 ft. apart.

(Continued on next page.)

Appendix (9), continued.

First case. The train is *not* moving.

The first sketch shows the case where the engineer has just blown the train whistle and the sound waves set out on their journey to the man's ear at the rate of about 263 waves per second. (They're just starting so you can't see the waves yet.)

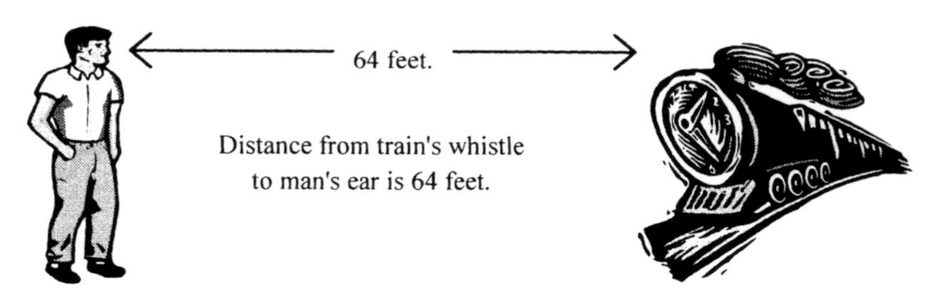

The next sketch shows the situation exactly *one sixteenth of a second later* when the *first* wave has just reached the man's ear.

By this time at 263 waves per second a total of *sixteen* (rounded off to the nearest whole number) waves have left the train's whistle and now fill the 64 ft. space between the whistle and the man's ear.

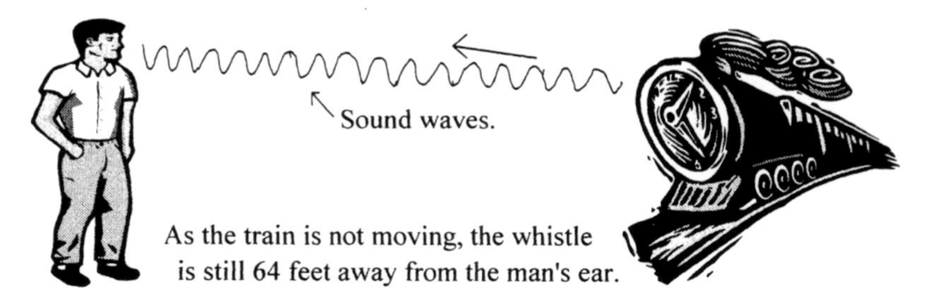

You'll see that *sixteen* waves have been emitted so far which fill the entire 64 ft. between the train's whistle and the man's ear.

Remember the waves *at this frequency* are 4 ft. long, so that 16 waves each 4 ft. long will just fill the 64 ft. space.

Appendix (9), continued.

<u>Second case. The train *is* moving</u>

Now we see the same situation as on the previous page, *except that now the train is moving.* Just as before, the *first* sketch shows the train whistle 64 feet from the man standing beside the track.

The engineer gives a mighty blast on the whistle, just like in the first case, and the first wave is once more setting out on its journey to the man's ear.

64 feet.

Just as before, as the engineer blows the whistle, the distance from the whistle to man's ear is 64 ft. Only this time the train and the whistle are moving!

The *second* sketch once again shows things *one sixteenth of a second later*, when the first wave *has just reached* the man's ear.

But while all this was happening the train (and whistle) were moving!

Sound waves.

This time the train's moving, and the distance from the whistle to the man's ear is only 58½ feet.

The train has moved nearer to the listener.

Appendix (9), continued.

We know it will take one sixteenth of a second for the first wave to reach the man's ear and, just as before, sixteen waves (to the nearest whole number) will have been emitted during this time.

Only this time the train is moving. So that by the time the first wave reaches the man's ear *the train has moved nearer to the man.*

If the train is traveling at 60 miles per hour it will have moved 5½ feet nearer to the man during this one sixteenth of a second.

We said that during this one sixteenth of a second the same number of waves had been emitted. *But now they're squeezed into only 58½ feet instead of the 64 feet they occupied before*. The waves have been all scrunched up together.

So if the waves are squeezed into only 58½ feet instead of the 64 feet they occupied previously, they'll be *shorter* than they would have been had the train not been moving. And a shorter wavelength means a *higher frequency*, which means the man will hear a *higher pitch note* as the train (and the whistle) are coming towards him.

Appendix (10).

On pages 52 to 54 we examined the case of a railroad coach which was traveling along the track at half the speed of light. The coach had two doors, one at each end of the coach, and located next to each door was a light-sensitive relay.

In the middle of the coach was a light bulb. The light bulb was located exactly midway between the two doors. When the light bulb flashed the light would travel outwards towards the two doors, and when it reached either relay this would cause its respective door to open.

To test the doors a man *inside the coach* pushes a button to make the light bulb flash. He sees the light moving towards each of the relays that open the doors.

How fast was the light traveling? To the man inside the coach it was traveling at 186,000 miles per second. Light *always* travels at 186,000 miles per second *relative to the person observing it.* (That's what the whole first part of this book was about.)

As the man is *inside* the coach he sees the light as traveling at 186,000 miles per second *relative to everything else in the coach.*

The doors are each *the same distance* from the light bulb and the light is traveling at *the same speed* towards each of the doors. It's clear the light is going to reach each of the doors *at the same time!* All this as seen by *(or relative to)* the man *inside* the coach. He will see both doors open at the exact same instant!

It would be the same no matter how fast the coach was moving relative to the track. Or even if the coach wasn't moving at all, relative to the track. It's like the case of the man and his wife on the cruise ship (page 18). The only way the man inside the coach would know the coach was moving at all would be by looking out of the window to some passing reference point.

All this is exactly what one would expect. Common sense tells us both doors are going to open at the same instant.

The trouble is there's another man sitting on an embankment beside the railroad track. As the man inside the coach is pushing the button to make the light bulb flash the coach is about level with the man on the embankment. (It doesn't have to be exactly level with him, just so long as he can see the doors.)

Appendix (10), continued.

The question is now about the man on the embankment. As the coach is rushing past him, what does *he* see?

Here's a diagram of the coach and a man on the embankment.

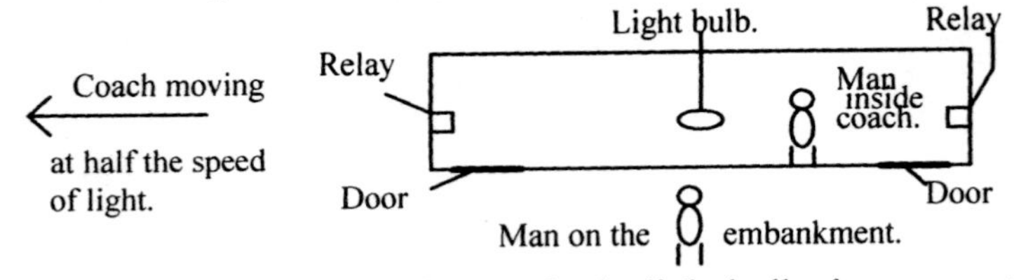

The above sketch shows the coach, the light bulb, the man on the embankment, and everything we talked about on the previous page.

At this instant the man inside the coach pushes a button to make the light bulb flash. (The beam of light can't be seen yet because the flash has *only just this second occurred*, and the light hasn't had time to emerge from the light bulb.)

Next are two separate sketches, the first like the one above, the other showing *only* the point from which the light sets out (which is where the light bulb was at the moment it flashed).

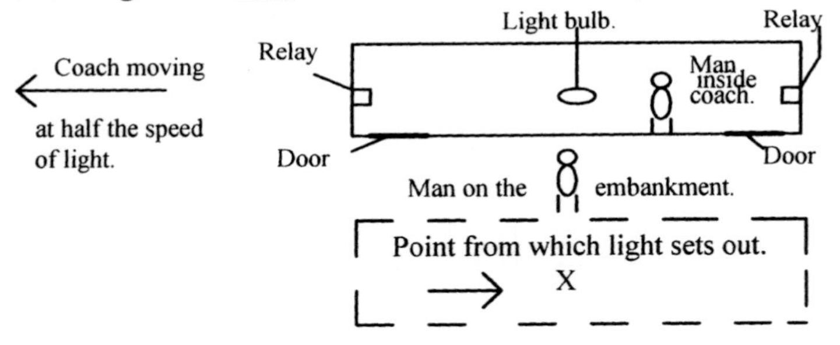

We've separated the coach and light bulb, *which we all know are moving*, from the point where the light started out which, *to the man on the embankment,* is *not* moving (relative to *him*).

The X in the middle of the outline of the coach above represents *the point from which the light sets out* on its way to the relays. (At the instant of the flash this is, of course, the light bulb.)

What we're concerned with now is exactly what *the man on the embankment* is going to see as the coach rushes by. Forget about the man *inside* the coach. (We know what *he's* going to see!)

Appendix (10), continued.

Through the windows of the coach the man on the embankment is going to see the flash of light. Then he's going to see the light spreading out from the point where the flash occurred.

What do we know about the speed of light? We know that light always travels at 186,000 miles per second relative to the observer. *And the observer in this case is the man on the embankment!*

The man on the embankment sees the light spreading forwards in the direction the coach is traveling, and backwards in the direction from which the coach has just come, *at the same speed relative to* *him* *in each direction!*

The man on the embankment is not moving *relative to the rail-road track.* So if the light is traveling at 186,000 miles per second *relative to the man on the embankment*, it must also be traveling at 186,000 miles per second *relative to the railroad track!*

After ten nanoseconds (ten billionths of a second) the light will have moved about ten feet in each direction *away from the point where the flash occurred.* (The light hasn't had time to reach the doors yet and still has a good way to go.)

Let's see how far the flash of light has gone.

⟵ x ⟶

The light has spread out *ten feet* in all directions *from the point where it started out.*

To the man watching from the embankment the light has moved ten feet along the track in the direction the coach is going, and ten feet along the track in the direction from which the coach has just come.

But what about the coach itself, and everything in it, including the doors, and the light bulb itself. *Because the coach is moving too!*

The coach, you will remember, is traveling at *half* the speed of light. So while the light has traveled ten feet the coach will have moved only five feet.

Now let's take a longer look at the coach, which is moving north along the track at half the speed of light.

Appendix (10), continued.

The coach, together with the light bulb, is moving north, *and to the man on the embankment no longer in any way influences how the light is traveling.* After ten nano-seconds, at half the speed of light, the coach has moved five feet.

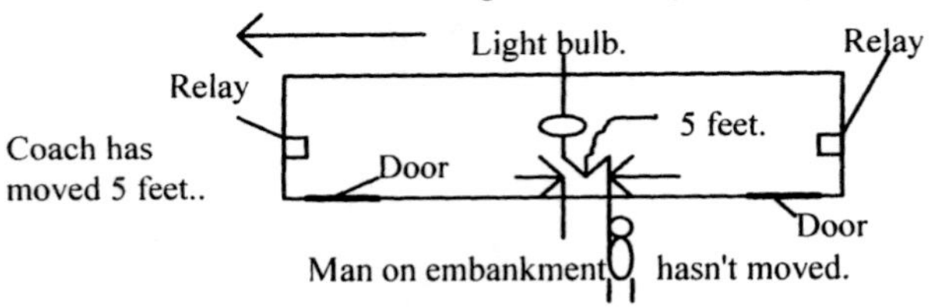

Next, putting the sketches side by side we see that the coach, together with the light bulb, has moved a distance of about five feet along the track. *But the point from which the light is spreading out has not moved.*

To the man on the embankment, the light is traveling at 186,000 miles per second *relative to him.* To *him* the light and the light bulb are going their separate ways!

Let's suppose that ten nanoseconds have passed since the light bulb flashed. To the man on the embankment the light has moved about ten feet north along the track, and ten feet south along the track. He also sees that the coach, together with the two doors, has moved five feet north along the track

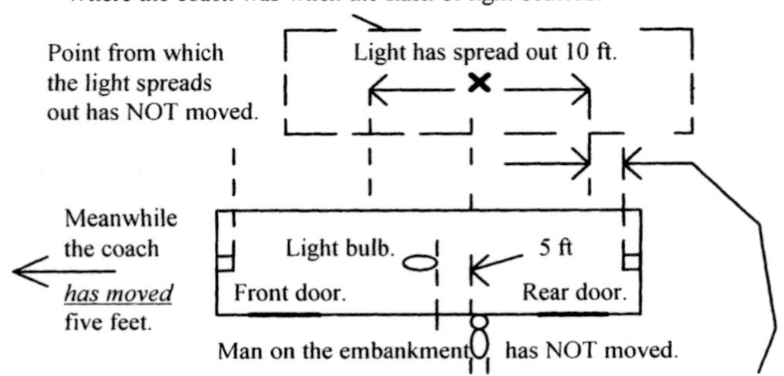

The above diagram demonstrates that, *as seen by the man on the embankment,* the light is now nearer to the rear door than to the front door, and is going to get there sooner.

Appendix (10), continued.

If the part about there being relays on the doors bothers you, then instead of relays just imagine that it is *the light itself* that opens the doors. We can suggest this because light, when it strikes something, imparts *energy*, one photon of light imparting one quantum of energy (which is why solar panels produce electricity; see also penultimate paragraph on page 150). So that if we imagine the doors to be light and flimsy, then we'll say the light itself simply pushes them open.

So to sum up the salient points, the man on the embankment sees the light bulb flash in the middle of the coach as it rushes by.

He sees the light set out in all directions, including north and south towards the two doors.

Once the light bulb flashes, *to the man on the embankment* subsequent movement of the light bulb is not relevant. The light and the light bulb have gone their separate ways.

The man on the embankment sees the beam of light heading both north and south traveling at 186,000 miles per second relative to him. *And therefore relative to the railroad track.*

The light heading north is approaching the front door which is also traveling north* but at only 93,000 miles per second.

So the gap between the light and the front door is closing at the rate of 93,000 miles per second. (the 186,000 miles per second that the *light* is traveling minus the 93,000 miles per second that the *door* is traveling in the same direction).

The man on the embankment also sees the beam of light heading south traveling at 186,000 miles per second relative to the railroad track. It is heading towards the rear door of the coach.

But in the case of the *rear* door it is rushing north *towards* the light at 93,000 miles per second. The train is traveling at half the speed of light, remember. (And it's traveling north, see page 53.)

So the beam of light and the rear door are approaching each other at the combined rate of 279,000 miles per second.

This is three times faster than the speed at which the light is approaching the front door!

Since both relays were originally the same distance from the light bulb when it flashed, the light is going to reach the rear door before it reaches the front door,

Therefore, as seen by the man on the embankment, the rear door is going to open a tiny fraction of a second before the front door.

Appendix (11). Force and Acceleration.

Imagine a friend's car has a flat battery and he asks you to give him a push. Luckily the car is equipped with standard transmission, so that the engine can be started by engaging a forward* gear then letting out the clutch once the vehicle is going fast enough.

Let's say it's a small car that weighs 2,000 pounds. Remember this is its *weight*, because we found out (page 38) that it's *mass*, not weight, that matters when you're trying to accelerate something.

Let's further say that you're going to have to get the car moving at five miles an hour before it's going fast enough to start the engine. Five miles an hour is (about) 7·33 feet per second.

Finally, let's say you're a strong, healthy person able to push the car with a *force* of one hundred pounds.

First we need to find the rate of acceleration of the car. If F = ma then $a = F/m$. (We established this back on page 84.)

We already have two of the values, F and (so long as we use the conversion factor) m. So

$$a = \frac{100 \text{ lbs.}}{2{,}000 \text{ lbs.} \times \left\{\dfrac{\text{lb. sec}^2}{\text{lb. } 32 \cdot 2 \text{ ft.}}\right\}}\ ^{**}$$

It works out that the acceleration is about 1·61 feet per second *per second.* (Refer to page 50 if this is still confusing.)

So if the car is accelerating at the rate of 1·61 feet per second per second then it will take $7{\cdot}33/1{\cdot}61$ seconds, or 4 · 55 seconds, for the car to reach the required speed (if we ignore things like friction).

You can see that if the car had *twice the mass* then you'd have to push it *twice as hard*, with a force of 200 pounds instead of only one hundred pounds. *Or* the acceleration would be only half of what it was and you'd have to push the vehicle for twice as long.

* This assumes that you're going to push the vehicle *forward!* (We engineers have to be precise about these things.)

** On page 151 we explained why this conversion factor has to be used, due to the difference between *mass* and *weight*. In this case it goes to the bottom of the equation because it's the 2,000 lbs. *weight* that has to be converted to mass. (The 100 lbs in the numerator is a *force*, so there's no need to mess around with that.)

Appendix (11), continued.

The foregoing is an example of how Sir Isaac Newton's laws of motion are straight forward and easy to apply.

Unlike much of what's being taught in our hallowed halls of higher learning these days, engineering is a "hard" subject, in that it involves the imparting of *real* knowledge, and where the solution to a problem is always predicated either on the fundamental laws of the universe or on empirical data well established. Opinions don't count.

You can't fake it in engineering, you're either right or you're wrong!

Professor of Engineering.
(Not a person you'd want to mess with.)

Due to the uncompromising nature of the subject drop out rates can be high among engineering students, many of whom will transfer to less demanding courses consisting of a lot of made up stuff where it's impossible to fail and where, as Admiral Rickover (1900-1986) once put it when testifying before Congress, "*nobody is ever wrong, and everyone's opinion is as good as everyone else's.*" This achieves the politically desirable goal of placing a college degree within reach of the functionally illiterate, but at a cost of squandering the nation's educational resources on those the least likely to benefit, rather than concentrating such finite resources on our best and brightest, where the quest for knowledge would be the more profitably advanced.

The low status of engineers in the United States results in only about sixty-thousand entering the profession per year* while China, her principal competitor, graduates more than half a million.** Of those studying for post graduate degrees in engineering at colleges and universities in the USA, the majority are either immigrants, or students from other countries, mostly from Asia.

* We do, however, produce vast number of lawyers, who contribute nothing to the national wealth while consuming far too much of it!

** Some authorities put this figure at six hundred thousand. China is also said to produce more *English-speaking* engineers than does the United States. Of all the people in the world able to speak English, more than half of them live in either China or India!

Appendix (12). The Chemical Elements.

There exist today nineteen, and possibly twenty-five* elements that do not, in any significant quantity, exist in nature. These are:

#	Element	Symbol	#	Element	Symbol
# 93	Neptunium	(Np)	# 103	Lawrencium	(Lr)
# 94	Plutonium**	(Pu)	# 104	Rutherfordium	(Rf)
# 95	Americium	(Am)	# 105	Dubnium ***	(Db)
# 96	Curium	(Cm)	# 106	Seaborgium	(Sg)
# 97	Berkelium	(Bk)	# 107	Bohrium	(Bh)
# 98	Californium	(Cf)	# 108	Hassium	(Hs)
# 99	Einsteinium	(Es)	# 109	Meitnerium	(Mt)
# 100	Fermium	(Fm)	# 110	Darmstadt	(Ds)
# 101	Mendelevium	(Md)	# 111	Roentgenium	(Rg)
# 102	Nobelium	(No)			

It is believed that many of these trans-uranium elements were present at the time the earth was formed but have since decayed. In the case of Neptunium, for example, the four and a half billion years that the earth has existed represents over two thousand times the 2·14 million year half-life of that element's longest-lived isotope.

If there'd been a million tons of Neptunium on earth when it was in its infancy, a mere ninety-one half-lifes would have been enough for radioactive decay to reduce this amount to just *a single atom!*

Nevertheless traces of trans-uranium elements can be detected in nature even today, but in such small quantities, perhaps less than *one kilogram per planet*, that in practice they have to be manufactured.

* Russian scientists claim to have produced elements 112 thru 116, also 118, though as of the end of 2006 this has yet to be confirmed.

** University of California researchers were able to independently produce this new heavier and fissionable element, one seemingly not found in nature, soon to be called plutonium (P-239) in which a chain reaction could also be sustained. This achievement was very wisely kept secret from everyone except for those on the Manhattan Project (which included a large number of spies for the Soviet Union) who were trying to produce an atomic bomb. Had it not been for this discovery of plutonium there'd have been only one bomb, a single uranium bomb, available in August of 1945. (See top of page 77.)

*** Prior to 1997 this new element had unofficially (and deservedly) been known as Hahnium, for Otto Hahn, as mentioned on page 181.

Appendix (12), continued.

On the previous page was a list of elements that supposedly do not exist in nature. Below is a list of those that do.

1 Hydrogen, H.
2 Helium, He.
3 Lithium, Li.
4 Beryllium, Be.
5 Boron, B.
6 Carbon, C.
7 Nitrogen, N.
8 Oxygen, O.
9 Fluorine, F.
10 Neon, Ne.
11 Sodium, Na.
12 Magnesium, Mg.
13 Aluminum, Al.
14 Silicon, Si.
15 Phosphorus, P.
16 Sulfur, S.
17 Chlorine, Cl.
18 Argon, Ar.
19 Potassium, K.
20 Calcium, Ca.
21 Scandium, Sc.
22 Titanium, Ti.
23 Vanadium, V.
24 Chromium, Cr.
25 Manganese, Mn.
26 Iron, Fe.
27 Cobalt, Co.
28 Nickel, Ni.
29 Copper, Cu.
30 Zinc, Zn.
31 Gallium. Ga
32 Germanium, Ge.
33 Arsenic, As.
34 Selenium, Se.
35 Bromine, Br.
36 Krypton, Kr.
37 Rubidium, Rb.
38 Strontium, Sr.
39 Yttrium, Y.
40 Zirconium, Zr.
41 Niobium, Nb.
42 Molybdenum, Mo.
43 Technetium, Tc.
44 Ruthenium, Ru.
45 Rhodium, Rh.
46 Palladium, Pd.
47 Silver, Ag.
48 Cadmium, Cd.
49 Indium, In.
50 Tin, Sn.
51 Antimony, Sb.
52 Tellurium, Te.
53 Iodine, I.
54 Xenon, Xe.
55 Cesium, Cs.
56 Barium, Ba.
57 Lanthanum, La.
58 Cerium, Ce.
59 Praseodymium, Pr.
60 Neodymium, Nd.
61 Promethium, Pm.
62 Samarium, Sm.
63 Europium, Eu.
64 Gadolinium. Gd.
65 Terbium, Tb.
66 Dysprosium, Dy.
67 Holmium, Ho.
68 Erbium, Er.
69 Thulium, Tm.
70 Ytterbium, Yb.
71 Lutetium, Lu.
72 Hafnium, Hf.
73 Tantalum, Ta.
74 Tungsten, W.
75 Rhenium, Re.
76 Osmium, Os.
77 Iridium, Ir.
78 Platinum, Pt.
79 Gold, Au.
80 Mercury, Hg.
81 Thallium, Tl.
82 Lead, Pb.
83 Bismuth, Bi.
84 Polonium, Po.
85 Astatine, At.
86 Radon, Rn.
87 Francium, Fr.
88 Radium, Ra.
89 Actinium, Ac.
90 Thorium, Th.
91 Protactinium, Pa.
92 Uranium, U.

Hydrogen is by far the most abundant element in the universe, and always has been.

Oxygen is the most abundant element on earth, as well as being one of, if not *the* most highly reactive of all elements.

Appendix (13). On the Mathematical Probabilities of the Existence of Life Elsewhere.

It is the biggest question facing the human race today. Is there life anywhere else in the universe, other than here on earth?*

Most scientists seem to be convinced there's life elsewhere in the universe. And yet there's no evidence to support this.

By now we've been able to explore most of the solar system, to some extent at least, and yet nowhere does life appear to exist. Not even close! So why should anywhere else be any different?

Well it has to do with numbers. There are so many stars, their numbers so far beyond our imagination, that almost *anything* must exist *somewhere* on a planet or moon of at least *one* of them! This is the argument. So then let's take a closer look at this numbers game.

In appendix (4) we showed how unlikely it was that a deck of fifty-two playing cards, arranged in numerical sequence and by suit, when thoroughly shuffled should, upon being dealt out, appear once more in their original order. The chance of this happening, we said, was less than one chance in ten to the sixty-eighth power (10^{68}).

The chance was so small because it was the *compounded improbability* of a large number of *individually* improbable events all occurring one after the other.

Now let's get back to the stars. There are thought to be between one hundred billion and two hundred billion galaxies, each galaxy with from one hundred billion to two hundred billion stars! That would put the total number of stars at somewhat more than ten to the twenty-second power (10^{22}). Perhaps 4×10^{22}.

Now that's a pretty big number! *But it's way short of the odds against a deck of cards being dealt out in any specific order!*

Suppose every one of these 10^{22} stars had a planet like earth, and that on each of these planets there lived six and a half billion people. (About the population of earth today.) If each one of these (close to 10^{32}) people all over the universe began dealing out cards, and kept on doing so for the rest of their lives, the odds would still be *trillions to one* against a deck of cards *anywhere* being dealt out in exactly the same order they were when the new decks were first opened.

Under this scenario the number of stars, their planets, and all the people living on them, is *inconsequential* compared to the *massive* improbability of the cards ever being dealt out in any specific order.

* What we mean here is life of a form that we could comprehend.

Appendix (13), continued.

So now let's talk about how improbable it is that life should exist at all! Because it seems that a lot of special conditions have to be present before a life-friendly planet like earth is possible. *And every one of these conditions appears in itself to be highly improbable.*

We said there are about 10^{22}stars. If the likelihood of conditions necessary for life to exist, namely a suitable planet or moon, were less than one chance in 10^{22}, then it's unlikely there'd be life anywhere else in the universe. The odds would simply be against it.

The next step, then, is to look at just exactly *how* improbable these necessary conditions are! Try and put a number on it!

For earth to have formed the way it has is unlikely. There's our moon for a start. No other body that we know of has a moon so large, both in mass *and* in relation to the planet that it orbits. Our home would have been radically different were it not for the moon being the way it is. For one thing it maintains the earth's axis at a constant angle relative to the plane of its (the earth's) orbital rotation. Without the moon's stabilizing effect the seasons would fluctuate wildly all over the planet, from burning desert one moment to frozen wasteland the next, and everything in between, making the establishment of life on earth highly unlikely. And the chances of our having just the *right kind* of moon could be one in *tens of thousands*. This is just one of a long list of unlikely conditions that appear to be necessary for a planet like ours to exist!

What about our magnetic field? Apart from earth, the only other terrestrial planet (at least in the solar system) with a bipolar magnetic field is Mercury, and that has only about one-sixth the strength of ours. (Other sources say it's a lot less than this.) It seems like just another one of those fortuitous circumstances! Yet if we didn't have a magnetic field much like the one we've got then our planet would be sterilized by lethal cosmic radiation (see page 28). Life would not be destroyed, for the simple reason that it would never have had a chance to get started here in the first place.

Liquid water is another thing thought to be essential for life (as we know it) to exist. After all it was in the ocean where life on earth first appeared! Even today 94% of all life on earth is aquatic! Yet with our great oceans of liquid water we appear to be unique. Earth, in fact, is the only planetary surface that we know of where materials in solid state and liquid state enjoy a stable coexistence!

Appendix (13), continued.

Then there's our oxygen level! It's just right! A bit higher and the first lightning strike would start a fire that no-one could put out, that would rage out of control until the last available molecules of carbon or oxygen on earth were consumed.

Then there's our location. We're *just the right distance* from the sun, in what scientists call the *habitable* zone. Any nearer to the sun it would be too hot. Any farther away from the sun and it would be too cold. *In any other than our present location we would either fry or freeze!* Some call it the Goldilocks effect, named for the nursery rhyme character who, upon entering the home of the three bears, was fortunate enough to find a bowl of porridge that was neither too hot, nor too cold. It was just right!

Although we're now finding planets of other stars, none of them appear to be anything like earth! Most are gaseous giants. But while the outer planets of our own solar system, Jupiter, Saturn, Uranus and Neptune, are also gaseous giants (certainly Jupiter and Saturn would fit this category) these newly discovered bodies are different. For one thing they're mostly in elliptical orbits that carry them so far out into space that temperatures must plunge to hundreds of degrees below zero, then so close to their parent stars that they'd be hotter than the sizzling side of Mercury facing the sun! Roughly circular orbits, like those of all eight of the planets in our solar system, have not been found elsewhere.

No other celestial body that we've ever been able to explore up to now offers even the *remotest* possibility of conditions favorable to the formation or the sustaining of life as we know it.

So if conditions here on earth are the result of a great number of circumstances, *each one of them in itself highly improbable*, then the *compounded* improbability of all of them occurring *together* could approach the improbability of a deck of cards, after shuffling, being dealt out in any specific order.

The trouble is, in trying to calculate the odds, although we know a lot of unlikely conditions have to be present for life to exist, we don't know *how* many. Furthermore the mathematical probabilities of the individual conditions existing cannot be calculated in any meaningful way with the imprecise data we have now! So let's make an assumption.

Appendix (13), continued.

Let's assume the chances of there being another planet like earth are *trillions* of *trillions* of *trillions* of times greater than the chance of a deck of cards being dealt out in any specific order. Let's say instead of one chance in 10^{68} it was as high as one chance in 10^{30}.

If the chances of there being another planet like earth were as high as one chance in 10^{30} that would mean that, even with more than 10^{22} other star systems where one could exist, the likelihood of there being another planet like ours *anywhere else in the universe* was still only one chance in a hundred million! Which would mean the odds on there being *life* elsewhere, life as we know it, were less than one chance in a hundred million! Which is for all practical purposes no chance at all.

To sum up then, there are so many stars the probability is high that at least one of them, *somewhere*, must have a planet similar to earth. On the other hand, for there to be a planet like earth requires so many unusual circumstances that the *compounded* improbability of all of them occurring together could overwhelm the *probability* generated by the enormous number of places where such a planet could exist. There are just too many unknowns for anyone right now to come up with a definitive answer to which way it is!

But the experts *want* for there to be life elsewhere. It's obvious, they say. Just look how many stars there are. And don't confuse us with the lack of facts. But we've been down this road before.

Back in the eighteenth century everyone thought there had to be a huge undiscovered continent located somewhere in the temperate regions of the southern hemisphere. This too was obvious. It even had a name. Terra Australis Incognita, they called it, from the Latin *Terra* meaning land, *Australis* meaning to the south, and *Incognitus* meaning unknown. (We're not talking about present day Australia here.) Fifty million people lived in this new land, it was estimated. The population of the American colonies at the time was only about three million. It would make a nice addition to the British Empire.

It must be there, the experts said, to balance the great continents to the north. For land to be distributed so unevenly made no sense! This was a view expounded principally by a gentleman by the name of Alexander Dalrymple. He was far from alone, however, for most of the leading members of the scientific community at the time were also convinced of the existence of this mythical great south land.

Appendix (13), continued.

Now these weren't a bunch of idiots. We're talking about some of the finest brains in Europe here. Fellows of the prestigious Royal Society, presidents of which at one time or another included the great Sir Isaac Newton and Sir Joseph Banks (the same Banks who was later to sponsor the famous and ill-fated voyage of the Bounty, under the command of his protégé, that extraordinarily fine seaman, the much maligned Lieutenant, later Rear-Admiral, William Bligh).

"*It doesn't exist*," muttered Captain Cook (Bligh's mentor) who, on a subsequent mission (and apart from the Hawaiians themselves of course) was the first to discover Hawaii. In a fruitless search for Dalrymple's great south land he'd sailed farther south than any man had gone before.

"*That it does not exist has not been proven to my satisfaction*," replied the Earl of Sandwich, First Lord of the Admiralty and after whom (to keep on his right side) Cook was later to name his newly discovered islands.* (More on Lord Sandwich in a moment.)

So off went the intrepid captain once again, heading south with two new ships this time, the Resolution and the Adventure, the latter vessel under the subordinate command of a Lieutenant Furneaux.

On his first voyage Cook had only a single ship, the Endeavour, which was nearly lost on the great barrier reef. Hence the decision to send two ships on future expeditions. (Unfortunately the vessels became separated, ultimately returning to England independently, so that the purpose of having two ships in the first place was defeated.)

* For a long time the Hawaiian Islands were known as the Sandwich Islands. Britain, in the person of a naval officer, Captain Thomas, claimed them for a while in the eighteen-hundreds, then voluntarily, for some reason, handed sovereignty back to the confused islanders. The Hawaiian flag, designed by Thomas and currently the state flag, features a Union Jack. Thomas Square, a small park located between Berretania and King Streets in Honolulu, is (presumably) named for the vacillating Victorian.

Appendix (13), continued.

And in the end, of course, Captain Cook (the son of a migrant farm worker, and almost entirely self taught) was right.

Now it is not the purpose of this appendix to persuade people that life does not exist elsewhere, but to caution against assumptions born of wishful thinking. Conjecture must be tested against fact.

It could be that our beautiful blue planet is unique! Or perhaps not. Nobody knows. Maybe one day we'll find out. Who can tell? But we need some *evidence* before we jump to conclusions.

And since we promised that we'd tell you more about the Earl of Sandwich (named for a place in England) it was another holder of that title who was responsible for the name becoming a household word!

It being his Lordship's custom to spend the winter in Monte Carlo, that fashionable watering hole for the well-bred and well-heeled during La Belle Epoque, and reluctant to leave the gaming tables when hungry, this scion of the upper class, who probably never did a days work in his life, would have the waiter bring him a piece of meat between two slices of bread. That way he could eat with one hand while gambling away the family fortune with the other. The idea caught on, and it soon became customary for a player on a winning streak, when ordering this forerunner of the Big Mac, to ask the waiter to bring him "*one of those things Lord Sandwich has*." The waiter would retire to the kitchen. "*Another sandwich*," he'd tell the cook!

But perhaps the best story concerns a successor to that title who found himself on the losing end of an acrimonious and, unhappily for him, very public verbal exchange. In denouncing a member of the opposing political party Lord Sandwich had angrily declared the man to be a scoundrel, prophesying that he would "*either be hanged, or die of the pox.*" To which the man replied, "*That would depend on whether I embrace your Lordship's principles, or your Lordship's mistress.*" (Sorry, but we seem to have gotten away from the subject, don't we.)*

* John Wilkes (1727-1797) is said to have been the man with the witty retort that most of us would have only thought of afterwards. Whether or not that gentleman's eventual demise was occasioned by either of the circumstances so scathingly forecast by Lord Sandwich is something the author has not been able to establish.

Appendix (14). The Achievement of Nuclear Fission.

On page 75 we told you how in Berlin in 1938 two Germans, Otto Hahn and Fritz Strassmann, had been bombarding uranium nuclei with neutrons. Neutrons are subatomic particles with no electrical charge, and which had been identified by Sir James Chadwick of the Cavendish Laboratory in Cambridge, England in 1932. However, it had not been the object of the experiment to cause fission.

Uranium is the heaviest *naturally occurring* element, and was at that time the heaviest of all known elements, though there seemed to be no reason why even heavier ones should not exist. What the two Germans had been trying to do was to create just such an element.

It didn't seem to work. Even stranger was the fact that traces of (what was at first thought to be) radium were detected. Then they found it wasn't radium at all but *barium* that had appeared. Barium is an element not much more than *half* the atomic mass of uranium.

Where had this barium come from? *Was it possible the uranium nucleus had split in two?* Into barium and something else?*

That elements could be *altered* had already been established by Rutherford who in 1929 changed nitrogen into oxygen. (As neutrons had yet to be identified he'd used alpha particles, see page 75.) But oxygen and nitrogen are right next to one another in the periodic table. For an element to split into two *entirely different* elements, of *nowhere near* the original element's atomic number, was something that seems to have taken Hahn and Strassman by surprise.

But there was a problem! If a uranium nucleus were to split in two a great deal of energy would have to be present, in the order of about 200 million electron volts (200MeV)** and there appeared to be no place this energy could have come from. *Or was there?*

Measurements showed that over the course of the experiment the overall mass of the material involved had been *reduced.* The fission products that remained had a mass *less* than the mass of the material Hahn and Strassmann had originally started out with. It was less by an amount equal to about one fifth the mass of a proton.

* There is evidence that in 1934 Enrico Fermi (1901-1954) may have produced fission without knowing it.

** An electron-volt should not be confused with a *volt.* An *electron-volt* is a unit of *energy.* (It's about enough to make a grain of sand jump.) But that's not anything we need to go into here.

Appendix (14), continued.

Now by 1938 everyone had heard of Einstein's equation $E=mc^2$, and this showed that the energy equivalent of the missing bit of mass was just about the two hundred million electron volts (200 MeV) that would have to be present if the nucleus had in fact split in two!

In this manner, then, was the fission (as it has come to be called) of an atomic nucleus first achieved and its significance understood.

News of the achievement spread, and upon hearing details (such as how the release of neutrons might, if the fissionable mass were of the right shape/mass configuration, create a chain reaction) everyone realized that an atomic bomb was now a theoretical possibility.

A feverish effort was made to translate this laboratory finding into a useable weapon that would enable its possessor to prevail over his adversaries (most of the world at that time being either at war or on the brink thereof).

By 1945 the new weapon had been successfully tested, and was promptly deployed with dreadful effect against other human beings (see pages 77 & 82) subsequently ushering in an extended period of uneasy peace, known as the cold war, at the same time leading to the use of nuclear fission as an abundant and *less* environmentally harmful way of meeting man's exploding (no pun intended) energy demands. It will also probably lead to the premature extinction of the human race, and thereby hopefully allow nature to recover the precarious balance so nearly destroyed by man during his brief, selfish and irresponsible stewardship.

Despite a reluctance on the part of some to admit it, there seems to be little doubt that it was indeed Hahn and Strassmann who were the ones primarily responsible for this great scientific breakthrough. However, that fission of the uranium nucleus might be possible, even the concept of initiating a chain reaction, had been forecast earlier by Leo Szilard (1898-1964). It is not clear whether Hahn and Strassman were aware of this at the time of their experiment.

Others were involved in this line of research too. One of these was Lisa Meitner, a former colleague of Hahn. Although not even in the same country at the time fission was actually achieved, it was evidently Meitner who'd been the first to understand what had taken place, and realize that the energy necessarily involved had resulted from the splitting of the uranium nucleus. (Even so, it was Hahn and Strassman who actually *did* it, and that's what counts.)

Appendix (14), continued.

Otto Hahn was awarded the Nobel prize for chemistry (he being a chemist, not a physicist) in the citation for which the achievement of nuclear fission was specifically mentioned.

To their credit Hahn and Strassmann made no attempt to conceal their findings. One wonders what course history might have taken had they chosen to do so, and had the exploitation of this new source of energy, with its awesome potential as a weapon of war, been confined within the militarily ambitious Germany of that time.

As we said, Hahn received the (1944) Nobel prize in chemistry. Back then the prize was awarded only after extensive investigation into the significance of the achievement for which the award was being considered, and the integrity and worthiness of the candidate proposed. A background check was carried out, you might say. A survey of more recent recipients of the honor (particularly the peace prize) suggests this practice may have been discontinued.

Furthermore, although element #105 today is called Dubnium, for a while at least it had been known as Hahnium, named for Otto Hahn (see footnote page 171). Having an element named after you, even though in this case it turned out to be only temporarily, would suggest to most people a degree of eminence in the field of physics.

Also we should mention that the element Meitnerium (element #109) was named, and in this case permanently it seems, for that former colleague of Hahn we told you about on the previous page, the physicist Lisa Meitner, who'd been forced to leave Germany for political reasons prior to the unexpected and momentous conclusion of the experiment in which her earlier collaboration had represented a significant, though probably not decisive, contribution.

Neither Hahn nor Strassman received similar recognition, other than temporarily, for Hahn, as mentioned above.

Even though Hahn is known to have opposed the Hitler regime, the mere fact that he'd been working in Germany at that turbulent time in the country's affairs would no doubt have operated to his disadvantage. Perhaps to the extent of denying him the immortality which would have attached to his name occupying permanent place in the table of elements.

Plus there was the fact that Hahn was a chemist, and physicists were probably not too happy that it was a chemist who made one of the most important discoveries in the history of nuclear physics!